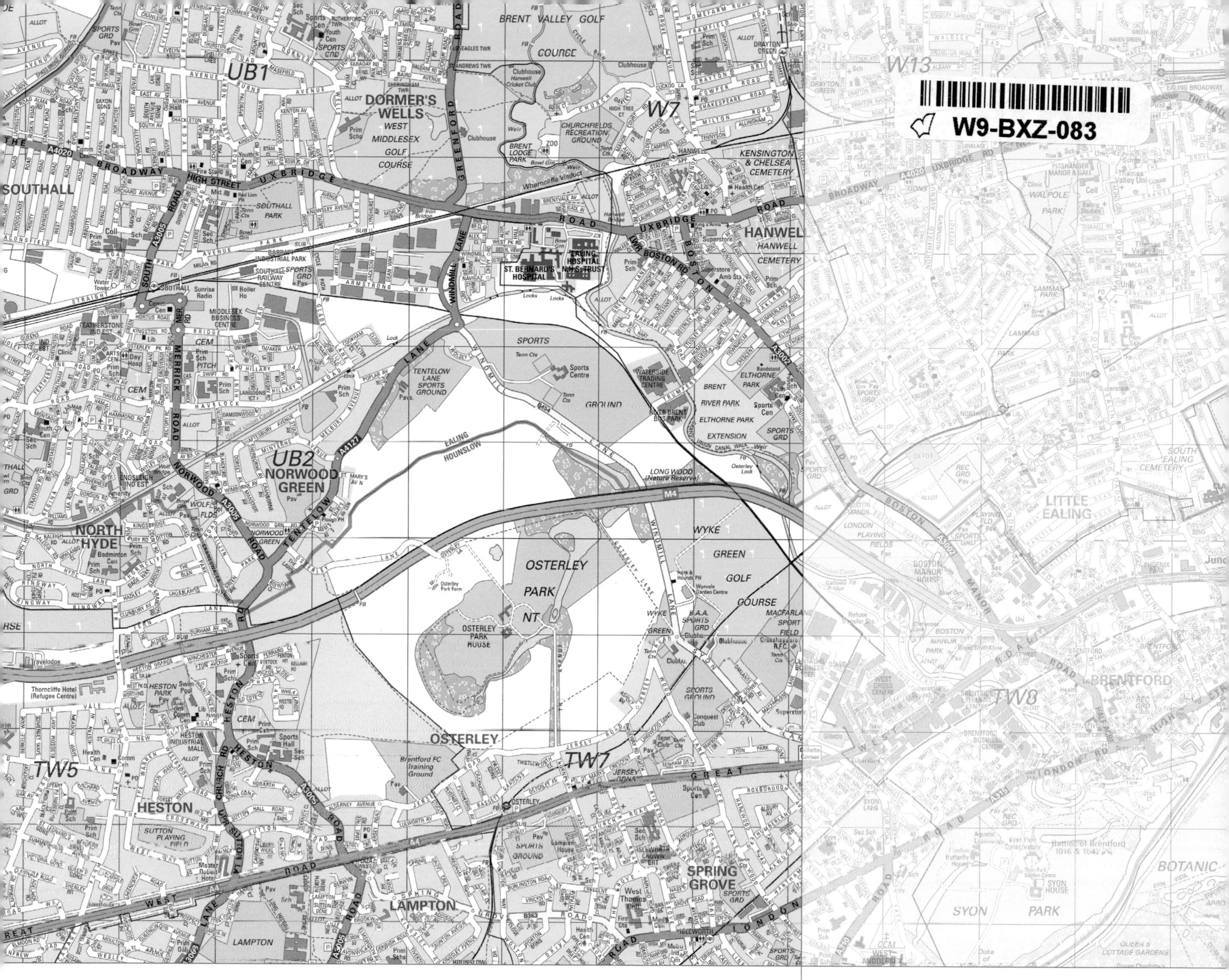

The fragile state of the world's economy, ecology, and security calls for innovative ways to help people process information and make wise decisions. Geographic information system (GIS) technology continues to play a critical role in that mission. *ESRI Map Book, Volume 24* reflects the creative applications of GIS—often in conjunction with other emerging technologies—on a global scale.

This year's sampling of finely crafted maps and displays attests to the versatility of GIS in government, conservation, public safety, transportation, utilities, tourism, and other human endeavors. Maps guide hikers to scenic treasures and lead tourists to fascinating discoveries. Maps document strategies for protecting vital habitat, distributing and conserving energy, monitoring crop productivity, and restoring areas ravaged by disaster.

GIS remains a rapidly evolving technology that, in the hands of professionals worldwide, serves as a foundation for visualizing, analyzing, and managing geographic data. I want to especially acknowledge the users who have allowed us to publish their maps in this book as extraordinary examples of using GIS to solve problems and reimagine the world.

Warm regards,

Jack Dangermond

Jack Dangermond

A Letter from Jack Dangermond

July 2009

Table of Contents

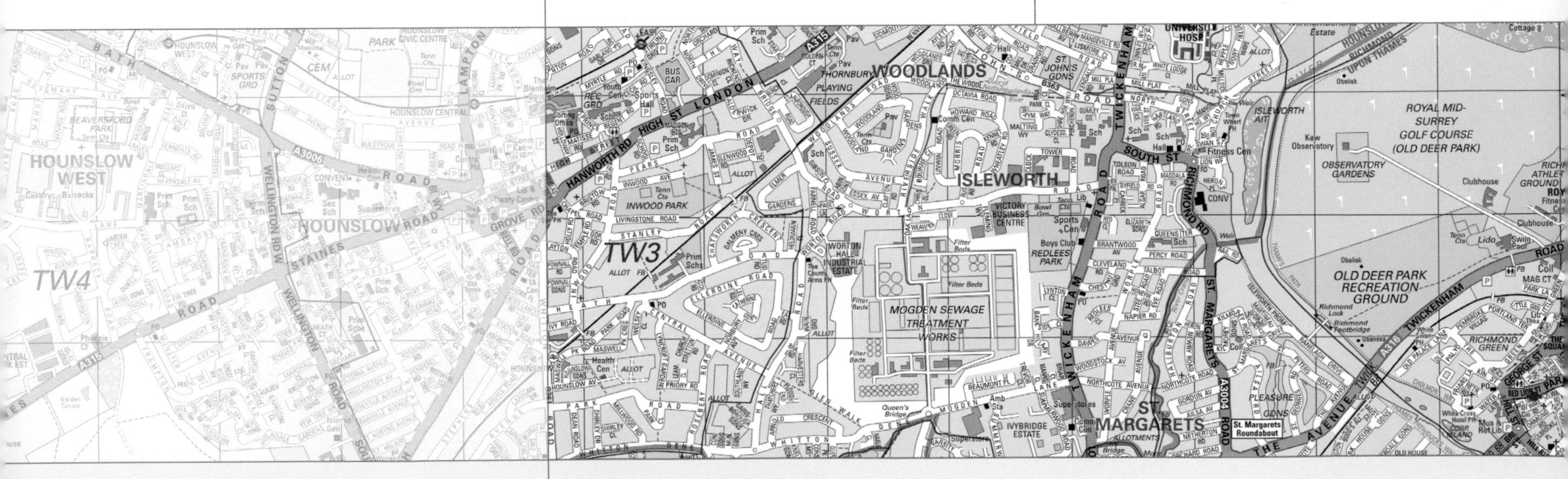

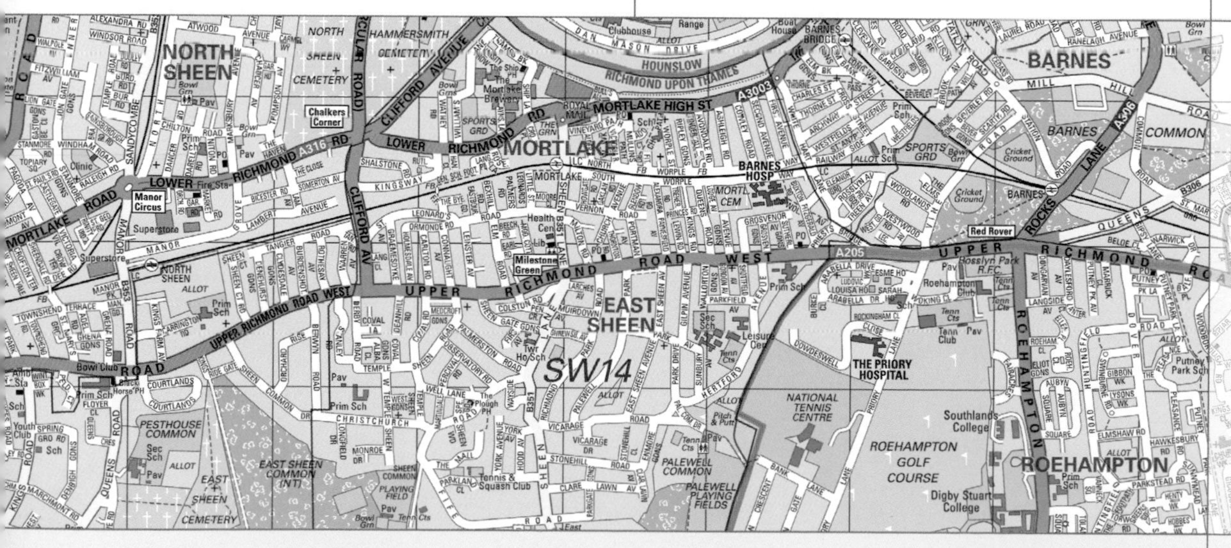

NORTH SHEEN
MORTLAKE
BARNES
EAST SHEEN
SW14
ROEHAMPTON
THE PRIORY HOSPITAL
ROEHAMPTON GOLF COURSE

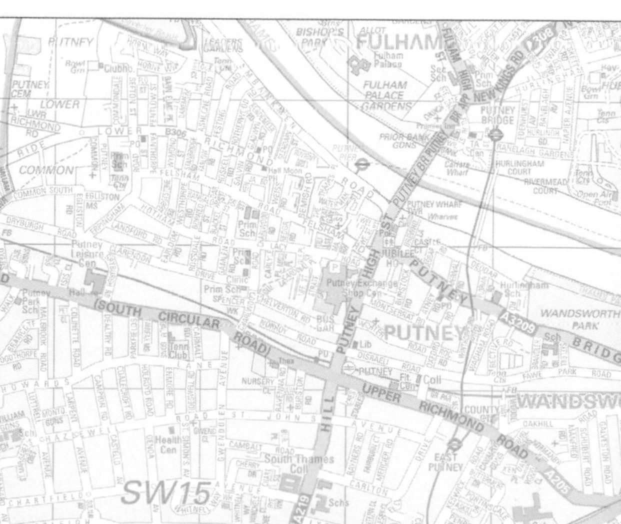

FULHAM
PUTNEY
SW15

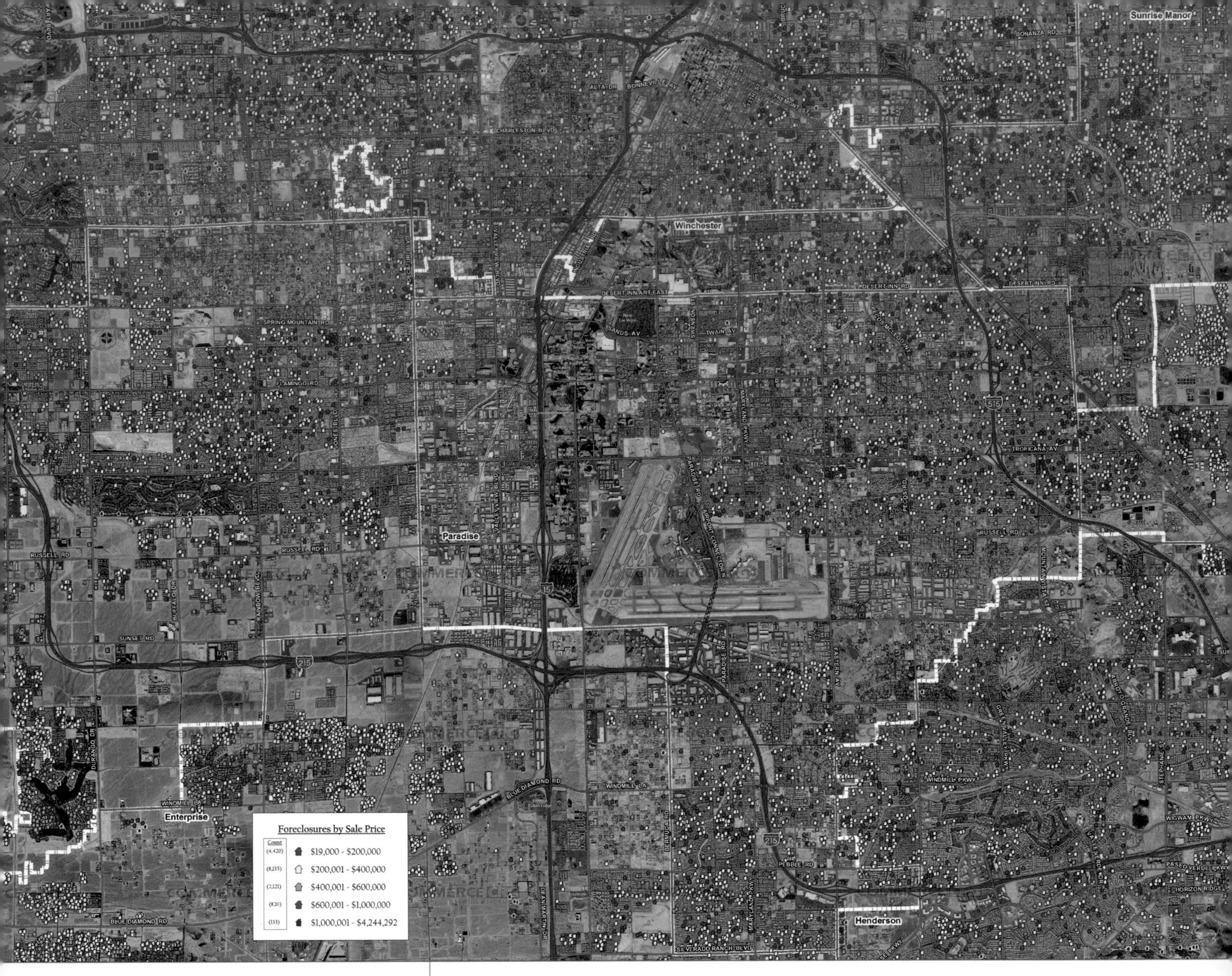

Commerce CRG
Salt Lake City, Utah, USA
By Ben Teran

Contact
Ben Teran
bteran@commercecrg.com

Software
ArcGIS Desktop 9.3

Printer
HP Designjet 4000

Data Sources
DigitalGlobe, First American CoreLogic

Commerce CRG, a commercial real estate brokerage company, created this map to analyze the struggling housing market in Las Vegas. The data is classified by the original sale price of the home. The foreclosure data is based on preforeclosure activity homes in danger of foreclosure. The map was created for the company's commercial real estate agents in Las Vegas to provide a visual representation of areas most affected by foreclosure.

Foreclosures generally have a negative effect on the commercial real estate market because residential slowdowns affect retail sales and industrial uses. Commerce CRG analyzed foreclosures as a way to identify landowners in the larger clustered developments and help them sell. The process also enabled agents to assist commercial property owners facing foreclosure.

Las Vegas Residential Foreclosures

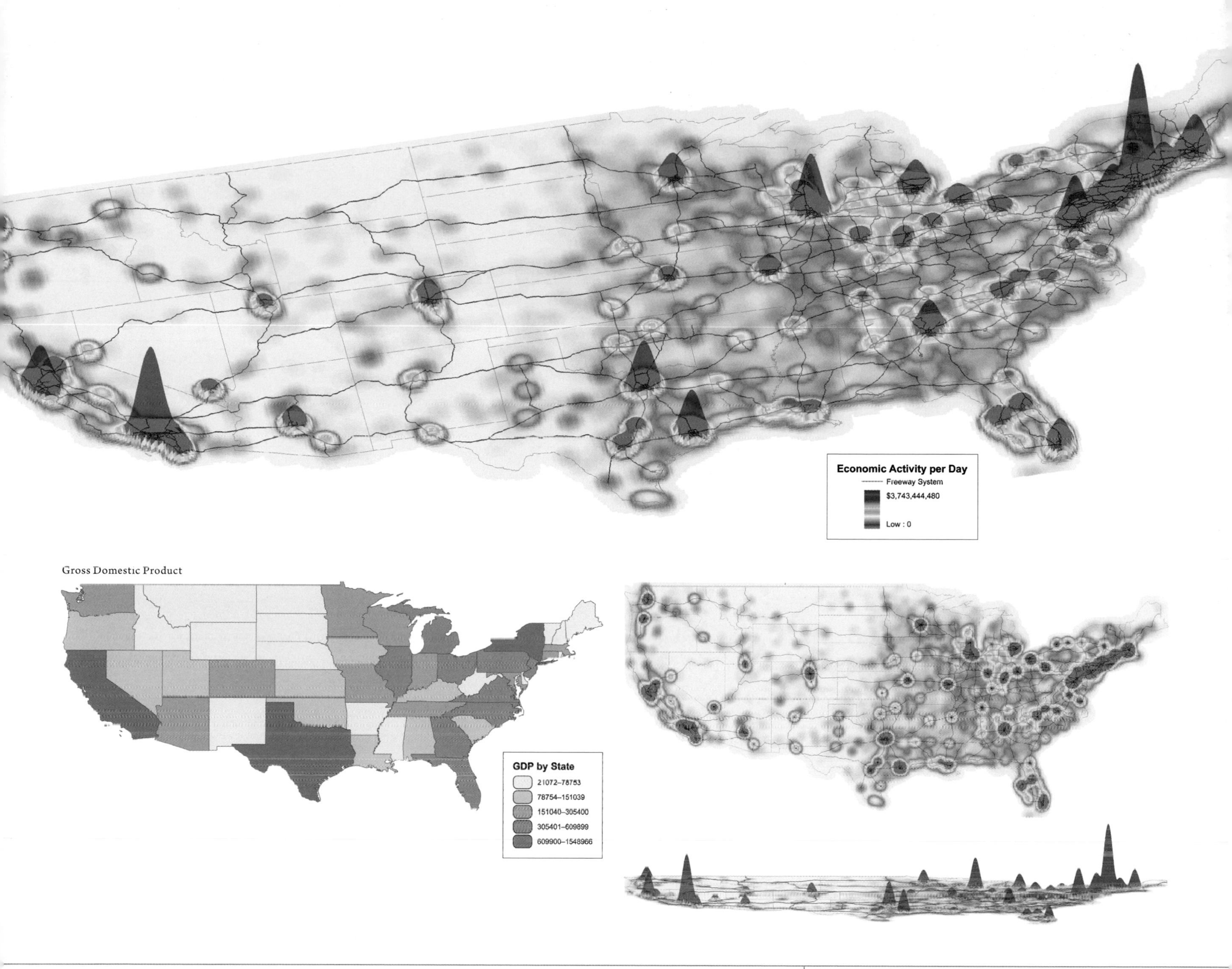

This map shows the distribution of economic activity across the continental United States using gross domestic product (GDP) produced per day as a measure. The scale of the economic activity is represented by the height and color from a 3D surface model. GDP is measured by using the employee data from Dun & Bradstreet and combining it with GDP by industry data from the Bureau of Economic Analysis (BEA).

Los Alamos National Laboratory uses this map to respond quickly to requests from federal and state agencies for economic impact analyses related to hazardous events. This map provides a unique perspective on economic activity that moves beyond tabular representations of economic data. This representation is intended to generate discussion and change perceptions about where economic activity is generated and the possible factors that explain geographic differences in economic activity.

Courtesy of Los Alamos National Laboratory.

Los Alamos National Laboratory
Los Alamos, New Mexico, USA
By Mary Ewers

Contact
Mary Ewers
mewers@lanl.gov

Software
ArcGIS Desktop, ArcGIS Spatial Analyst, ArcGIS 3D Analyst

Printer
HP Designjet 800ps

Data Sources
Dun & Bradstreet, BEA

U.S. Centers of Economic Activity (Continental United States)

Collins Bartholomew
Cheltenham, Gloucestershire, United Kingdom
By Mike Cottingham and Graham Gill

Contact
Graham Gill
Graham.Gill@harpercollins.co.uk

Software
ArcGIS Desktop 9.2, Maplex 3.5, Adobe Illustrator 10

Printer
Océ Arizona 250 GT

Data Source
Collins Bartholomew London Streets Database

This is a fully detailed street map at 1:20,000 scale covering over 4,000 square kilometers (1,545 square miles) around London. Originally designed in 1991, the specification has changed relatively little over the years, while the production technologies used have evolved massively.

This map specification is used in a variety of Collins Bartholomew-printed maps and atlases, and as raster data by organizations such as the Metropolitan Police, Visit London, local government departments, and UK Web mapping providers Multimap and Streetmap. This version is a 20-by-15-foot wall map.

Courtesy of Collins Bartholomew.

London Street Map

WHITECHAPEL
LIMEHOUSE
EC4
EC3
SHADWELL
WAPPING
POPLAR
E14
SE1
THE BOROUGH
ROTHERHITHE
BERMONDSEY
NEWINGTON
SE16
MILLWALL
SE17
WALWORTH
DEPTFORD
SE8
SE14
CAMBERWELL
PECKHAM
NEW CROSS GATE
NEW CROSS
SE15
ST JOHN'S
RIVER THAMES

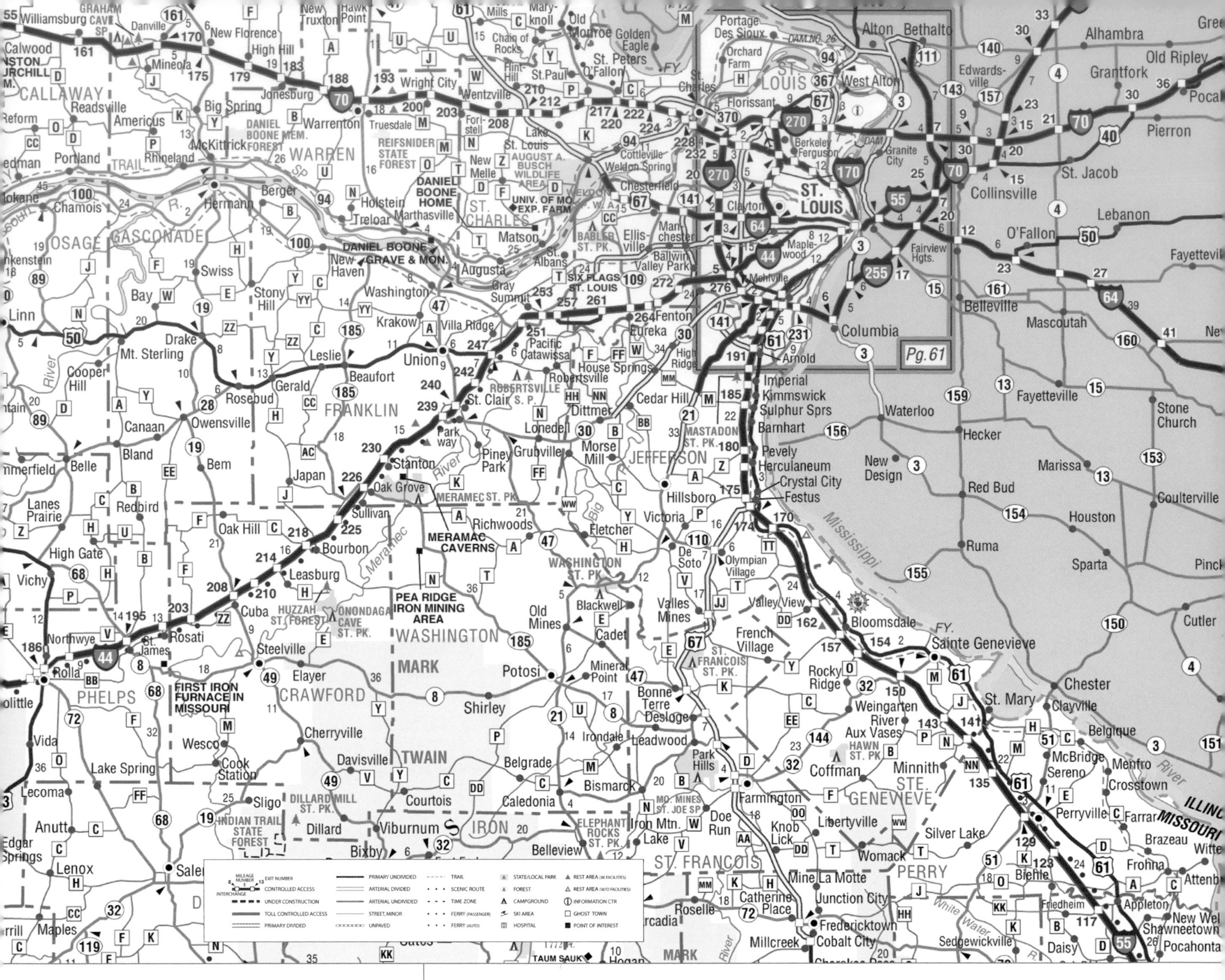

RMSI Private Limited
Noida, Uttar Pradesh, India
By Sridhar Devineni

Contact
Sridhar Devineni
sridhar.devineni@rmsi.com

Software
ArcGIS Desktop 9.2, ArcInfo NT, ArcObjects, ARC Macro Language (AML), AutoCAD Map, Adobe Illustrator 10.0, CorelDRAW 10, ARX, AutoLisp, C++, VB, and VBA

Printer
HP Designjet 5500ps

Data Sources
Various

This map is a representative sample of RMSI cartography services. RMSI, a global geospatial information and software services company, developed a cartographically styled geodatabase for street atlas maps using ArcGIS software. The system allowed for easy updates and also facilitated different types of products, including a pocket atlas, large print atlas, fold maps, and flip maps from a common source geodatabase.

RMSI develops innovative solutions that integrate geographic information with niche business applications and has expertise in natural disaster and climate change risk modeling. Its global client base covers such sectors as disaster management, insurance, agriculture and natural resources, land and property, government and multilateral funding agencies, telecom, and utilities.

Cartography Enabled Geodatabase for U.S. Cities

The Skagit County GIS Department is often asked to produce a general map that shows roads and political boundaries along with contour information. In the past, the U.S. Geological Survey's (USGS) 7.5-minute topographic maps were useful for this purpose. A previous project re-created the USGS 7.5-minute topographic maps using information that the GIS Department maintains and collects.

This new project took that product and made an 11-by-17-inch map book that could be distributed as an atlas. The maps were rescaled to 1:31,560 (one inch equals a half mile). This scale and size work well in vehicles and at people's desks, and has been popular with the community for hiking, hunting, and exploring the region. The atlas is printed on a high-quality laser jet printer and is sold for the printing cost.

Skagit County
Mount Vernon, Washington, USA
By Joshua Greenberg and Brian Young

Contact
Joshua Greenberg
joshg@co.skagit.wa.us

Software
ArcGIS Desktop, Adobe Photoshop, Microsoft Publisher

Printer
Icon 650 Laserprinter

Data Sources
Skagit County GIS, U.S. Geological Survey 10m data elevation model, National Landcover Database, Washington State Department of Natural Resources

City of Calgary
Calgary, Alberta, Canada
By Garnet Whyte and Wayne Scribbins

Contact
Audrey Stamm
audrey.stamm@calgary.ca

Software
ArcGIS Desktop 9.2 and MicroStation

Printer
HP Designjet 5500ps

Data Sources
The majority of the data on the map is from the City of Calgary's Digital Aerial Survey, which uses 1:5,000 aerial photography with a positional accuracy of better than 15 cm. The data is served up to the corporation through SDE.

This map was created many years ago using computer-aided design (CAD) software and became a mainstay on the walls of many Calgary businesses for determining building names, locations, and classifications as well as showing Calgary's Plus 15 walkway system within the downtown core area.

With the recent growth surge in Calgary, the map needed updating. To facilitate easier and quicker updating in the future, it was decided to convert the map to ArcGIS. The map was also enlarged to cover more area surrounding the core district.

Calgary Downtown Map

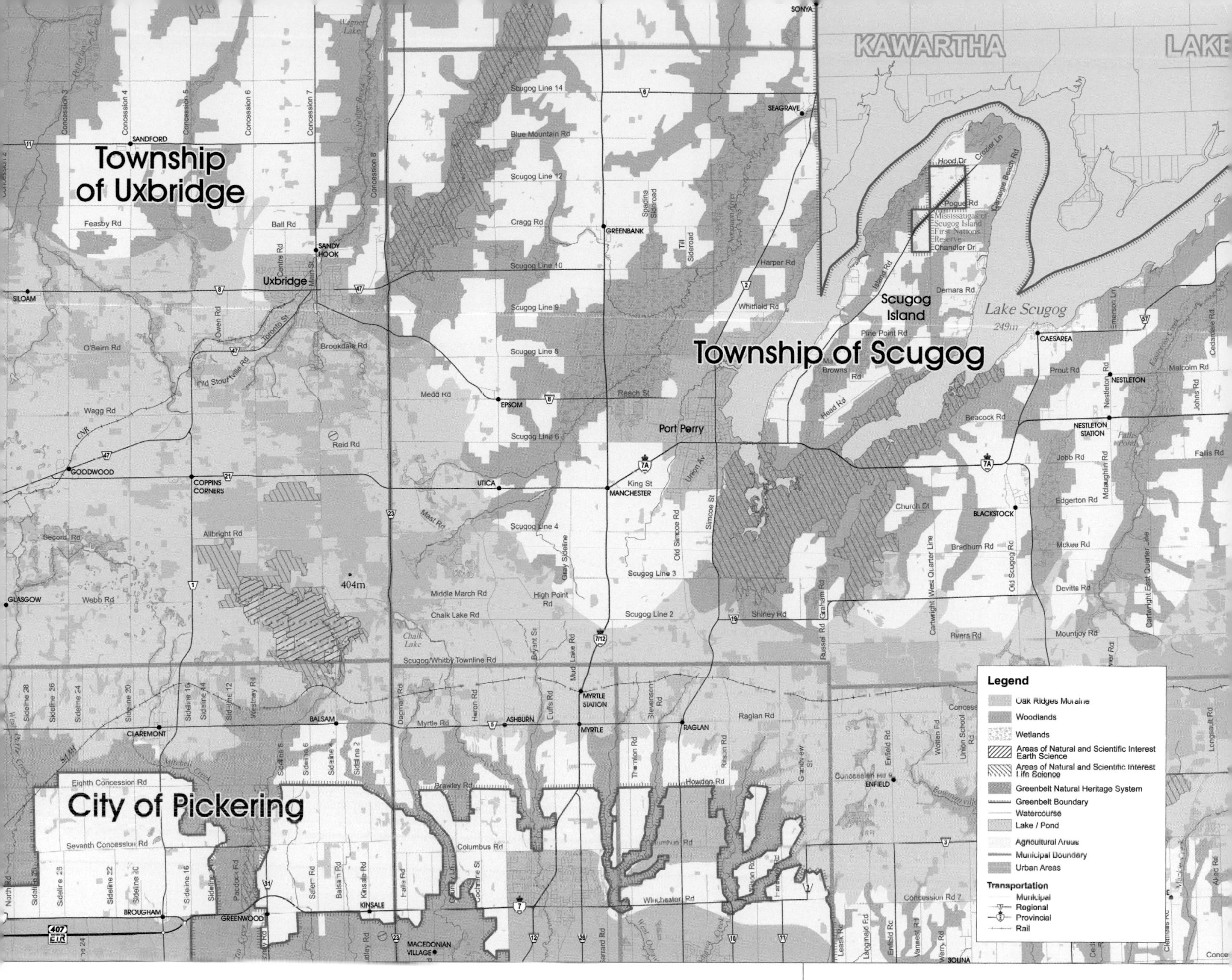

The Durham Environmental Advisory Committee uses this map to illustrate and educate the public about the natural features of the region under its mandate for community outreach. The map assists Durham residents in recognizing the role natural features play in defining the region, and provides basic information on how to appreciate and protect the region's natural features. This map has been distributed to secondary schools, libraries, and municipalities across the region.

The Regional Municipality of Durham Environmental Advisory Committee
Whitby, Ontario, Canada
By Lisa Hergott

Contact
Lisa Hergott
Lisa.Hergott@durham.ca

Software
ArcGIS Desktop 9.2

Printer
HP Designjet Z6100ps 42 in Photo PS3

Data Sources
The Regional Municipality of Durham, Ontario Ministry of Municipal Affairs and Housing, Ontario Ministry of Natural Resources, First Base Solutions

Region of Durham Natural Features Map

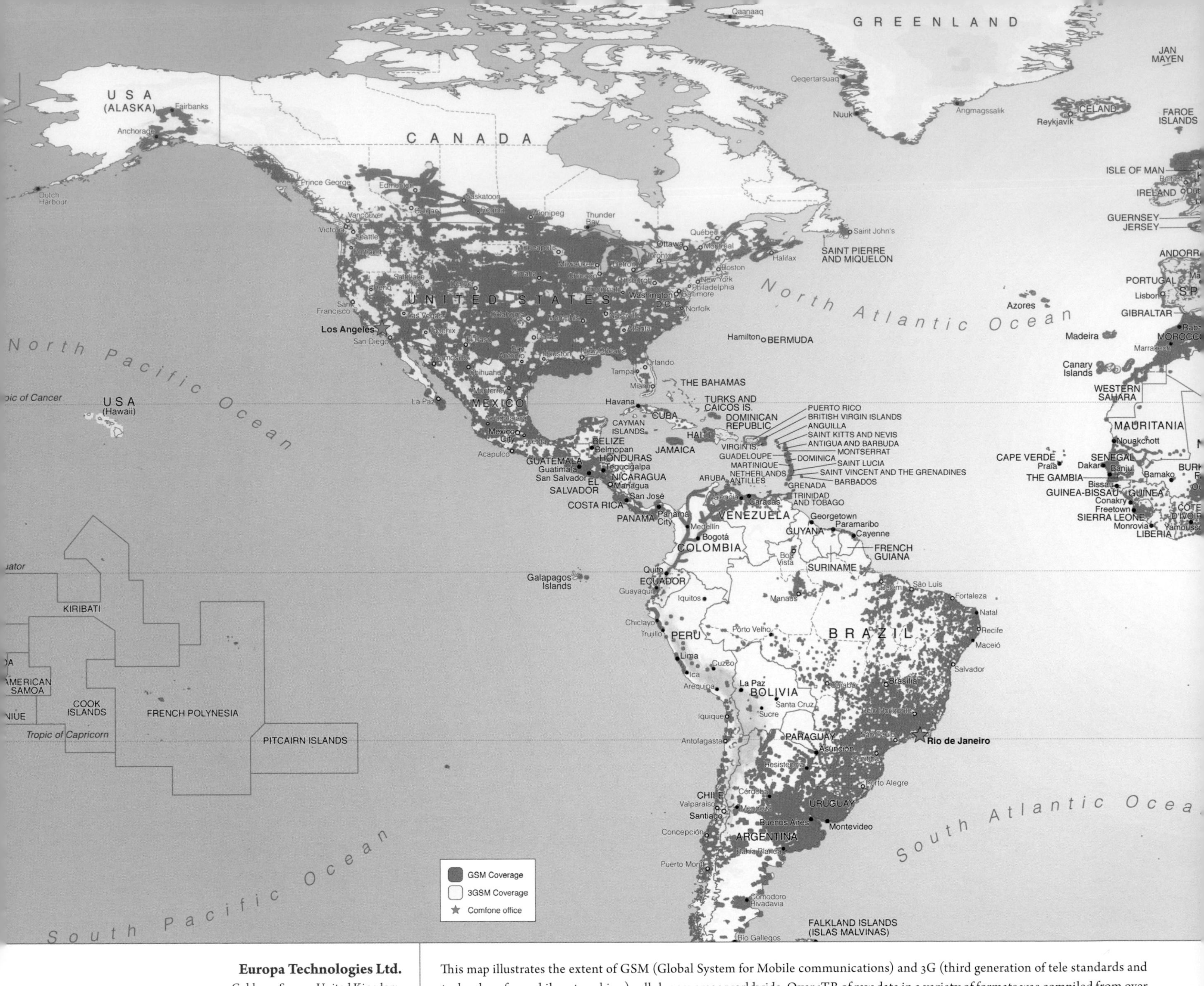

Europa Technologies Ltd.
Cobham, Surrey, United Kingdom
By Warren Vick

Contact
Warren Vick
wvick@europa.uk.com

Software
ArcGIS Desktop, MAPublisher, Adobe Creative Suite

Printer
Epson 7800

Data Source
Global Insight Plus

This map illustrates the extent of GSM (Global System for Mobile communications) and 3G (third generation of tele standards and technology for mobile networking) cellular coverage worldwide. Over 1TB of raw data in a variety of formats was compiled from over 500 network operators and homogenized with ArcGIS software. The world basemap is Global Insight Plus from Europa Technologies.

The publication is produced for the GSM Association, the global trade organization for the leading mobile telecommunications standard. Through sponsorship, over 7,000 copies will be printed and distributed at telecoms events, including Mobile World Congress in Barcelona.

GSM World Coverage 2009

R U S S I A
SWEDEN
FINLAND
Stockholm
Helsinki
Tallinn
ESTONIA
LATVIA
Riga
LITHUANIA
Vilnius
Malmo
Berlin
Warsaw
POLAND
BELARUS
Minsk
Moscow
Kiev
UKRAINE
MOLDOVA
Chisinau
SLOVAKIA
HUNGARY
AUSTRIA
ROMANIA
Bucharest
SERBIA
BULGARIA
CROATIA
ALBANIA
MACEDONIA
ITALY
Rome
GREECE
Athens
Ankara
TURKEY
GEORGIA
Tbilisi
ARMENIA
Yerevan
Baku
AZERBAIJAN
TURKMENISTAN
Ashgabat
UZBEKISTAN
Tashkent
Bishkek
KYRGYZSTAN
Dushanbe
TAJIKISTAN
Kashi
KAZAKHSTAN
Astana
Karaganda
Almaty
Urümqi
MONGOLIA
Uliastay
Ulaanbaatar
Norilsk
Igarka
Vorkuta
Salekhard
Arcangel
Surgut
Tiksi
Verkhoyansk
Yakutsk
Magadan
Okhotsk
Chita
Khabarovsk
Petropavlovsk-Kamchatskiy
Yuzhno-Sakhalinsk
Sapporo
Vladivostok
NORTH KOREA
Pyongyang
Seoul
SOUTH KOREA
JAPAN
Tokyo
Osaka
Beijing
Xining
C H I N A
Chengdu
Chongqing
Lhasa
Shanghai
Hangzhou
Taipei
TAIWAN
Macau SAR
Hong Kong SAR
Haikou
North Pacific Ocean
CYPRUS
SYRIA
LEBANON
Beirut
Damascus
ISRAEL
Jerusalem
JORDAN
IRAQ
Baghdad
Tehran
IRAN
Mashhad
Kabul
AFGHANISTAN
Kandahar
Islamabad
PAKISTAN
Karachi
New Delhi
NEPAL
Kathmandu
BHUTAN
Thimphu
BANGLADESH
Dhaka
INDIA
Mumbai
Hyderabad
Chennai
Bangalore
Mangalore
Cochin
Colombo
SRI LANKA
Malé
MALDIVES
KUWAIT
BAHRAIN
QATAR
Riyadh
SAUDI ARABIA
Abu Dhabi
UAE
Muscat
OMAN
YEMEN
Sana
Aden
Al Mukalla
Medina
Jeddah
Zahedan
Shiraz
Mandalay
MYANMAR
Rangoon
LAOS
Vientiane
Hanoi
THAILAND
Bangkok
CAMBODIA
Phnom Penh
Ho Chi Minh City
VIETNAM
Hue
Andaman and Nicobar Islands
Phuket
MALAYSIA
Kuala Lumpur
SINGAPORE
BRUNEI DARUSSALAM
Bandar Seri Begawan
Manila
PHILIPPINES
Cebu
Davao
Manado
Balikpapan
Padang
Palembang
Jakarta
Bandung
Surabaya
INDONESIA
Dili
TIMOR-LESTE
PAPUA NEW GUINEA
Port Moresby
NORTHERN MARIANA ISLANDS
GUAM
PALAU
FEDERATED STATES OF MICRONESIA
MARSHALL ISLANDS
NAURU
SOLOMON ISLANDS
VANUATU
CORAL SEA ISLANDS
NEW CALEDONIA
TUVA
Darwin
Broome
Cairns
Alice Springs
AUSTRALIA
Brisbane
Perth
Adelaide
Canberra
Sydney
Melbourne
Hobart
Auckland
NEW ZEALAND
Christchurch
Dunedin
Invercargill
South Pacific Ocean
Indian Ocean
MALTA
Tunis
Tripoli
Banghazi
LIBYA
Marzuq
EGYPT
Alexandria
Cairo
Aswan
Wadi Halfa
Port Sudan
CHAD
N'Djamena
Khartoum
SUDAN
Al Ubayyid
ERITREA
Asmara
DJIBOUTI
Adis Abeba
ETHIOPIA
SOMALIA
Mogadishu
CENTRAL AFRICAN REPUBLIC
Bangui
CAMEROON
Yaounde
PEOPLE'S REPUBLIC OF THE CONGO
GABON
Libreville
Kisangani
UGANDA
Kampala
KENYA
Nairobi
Mombasa
RWANDA
Kigali
BURUNDI
Bujumbura
DEMOCRATIC REPUBLIC OF THE CONGO
Kinshasa
Brazzaville
Kigoma
Dodoma
TANZANIA
Dar es Salaam
SEYCHELLES
Luanda
ANGOLA
Huambo
Lubumbashi
ZAMBIA
Lusaka
MALAWI
Lilongwe
COMOROS
MAYOTTE
Harare
ZIMBABWE
MOZAMBIQUE
Beira
Antananarivo
MADAGASCAR
MAURITIUS
REUNION
NAMIBIA
Windhoek
BOTSWANA
Gaborone
Pretoria
Maputo
Mbabane
SWAZILAND
Maseru
LESOTHO
Durban
SOUTH AFRICA
Cape Town
Port Elizabeth
East London

Projected Growth Impacts in 2050

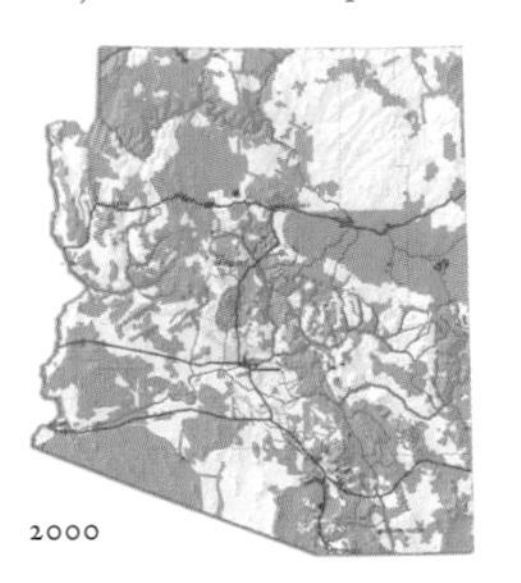

2000

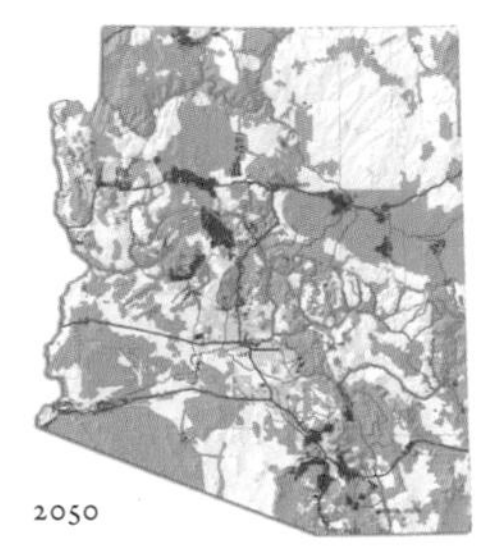

2050

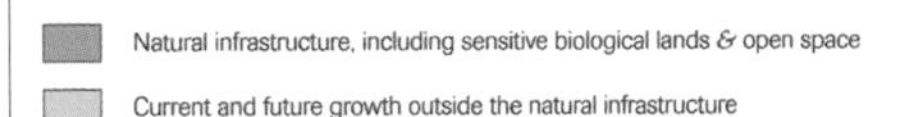

Alternative Scenarios of Arizona's Growth and Natural Infrastructure in 2050

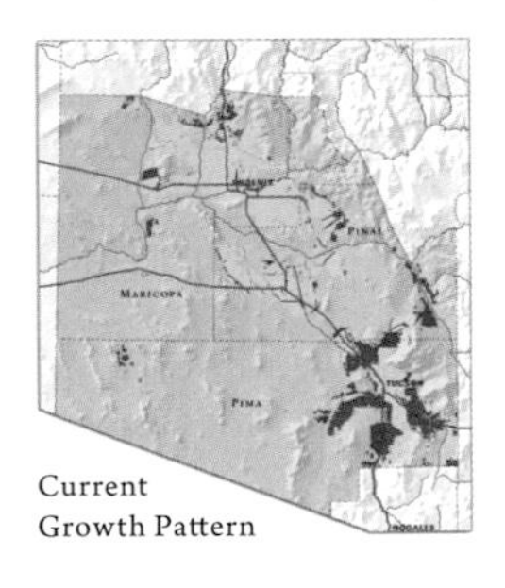

Current Growth Pattern

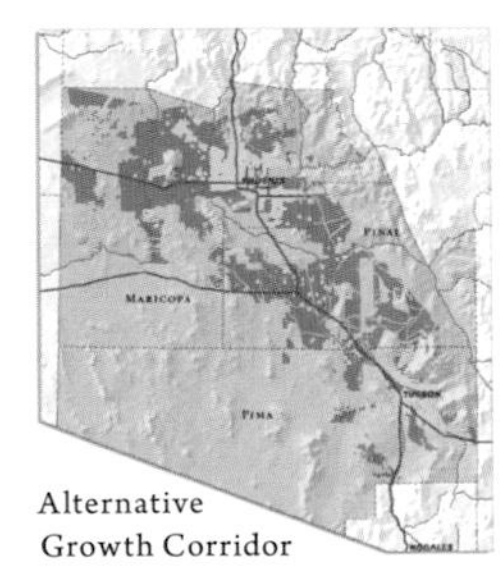

Alternative Growth Corridor

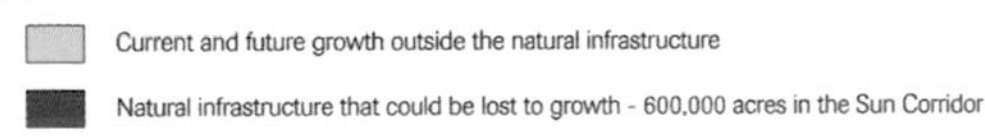

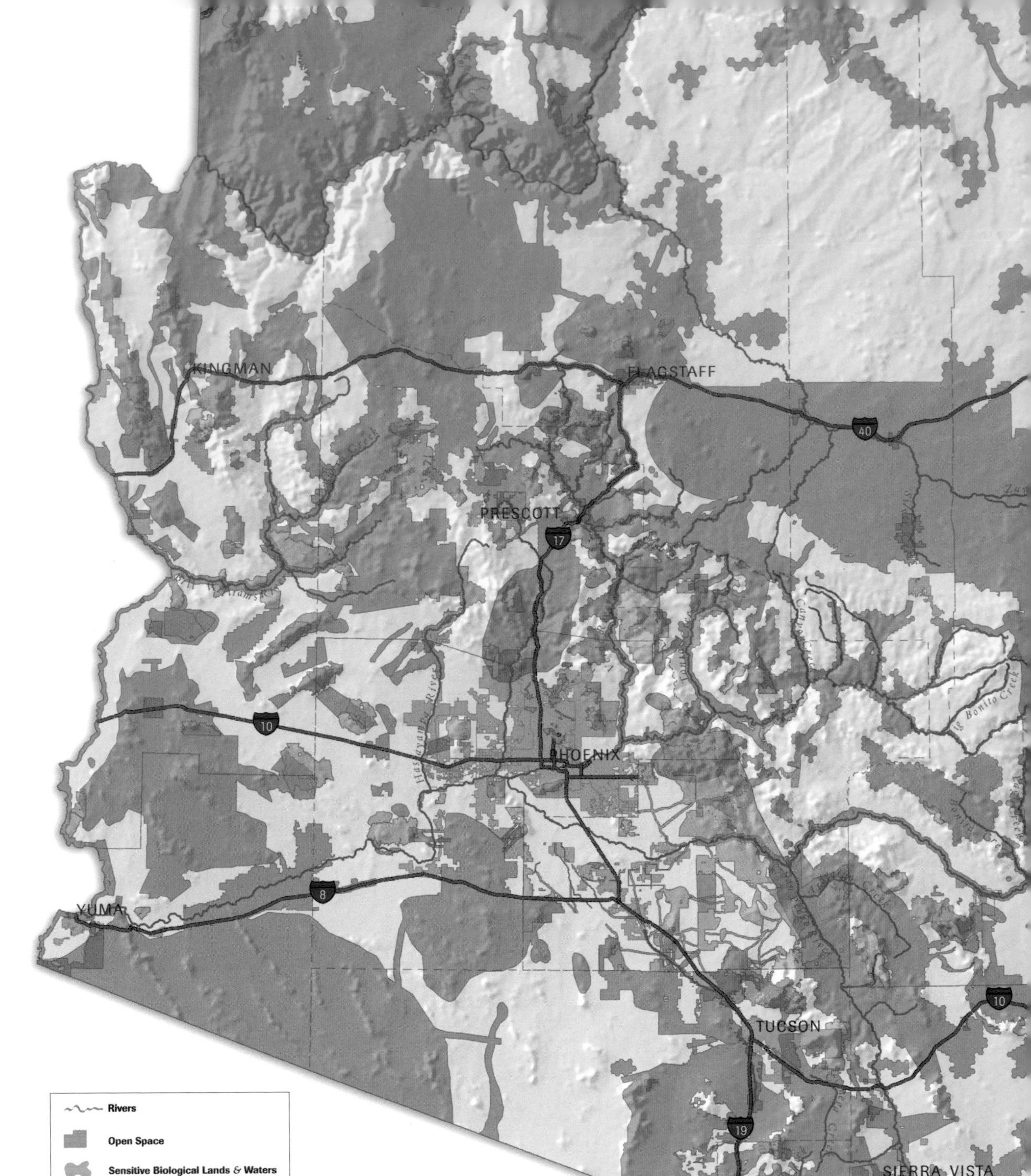

The Nature Conservancy
Tucson, Arizona, USA
By Dan Majka, Marcos Robles, and Rob Marshall

Contact
Dan Majka
dmajka@tnc.org

Software
ArcGIS Desktop

Printer
HP Designjet 800ps

Data Sources
The Nature Conservancy, state agencies

Arizona's natural infrastructure includes lands and waters that preserve the state's natural heritage and open space. The Nature Conservancy developed a natural infrastructure dataset by integrating twelve regional studies on wildlife habitat and open space, and used the dataset to understand the potential impacts of Arizona's future growth by 2050.

Results show that although Arizona's population is projected to double by 2050, its associated urban footprint may quadruple. If growth follows current projections, Arizona could lose nearly two million acres of natural infrastructure by 2050. This loss of desert, grassland, and forest habitat could jeopardize at least 120 species.

However, there are 2.7 million acres of undeveloped private and state lands outside of the natural infrastructure and within 30 miles of existing highways. Shifting projected development into these areas would minimize direct impacts to the natural infrastructure.

Courtesy of The Nature Conservancy.

Arizona's Natural Infrastructure

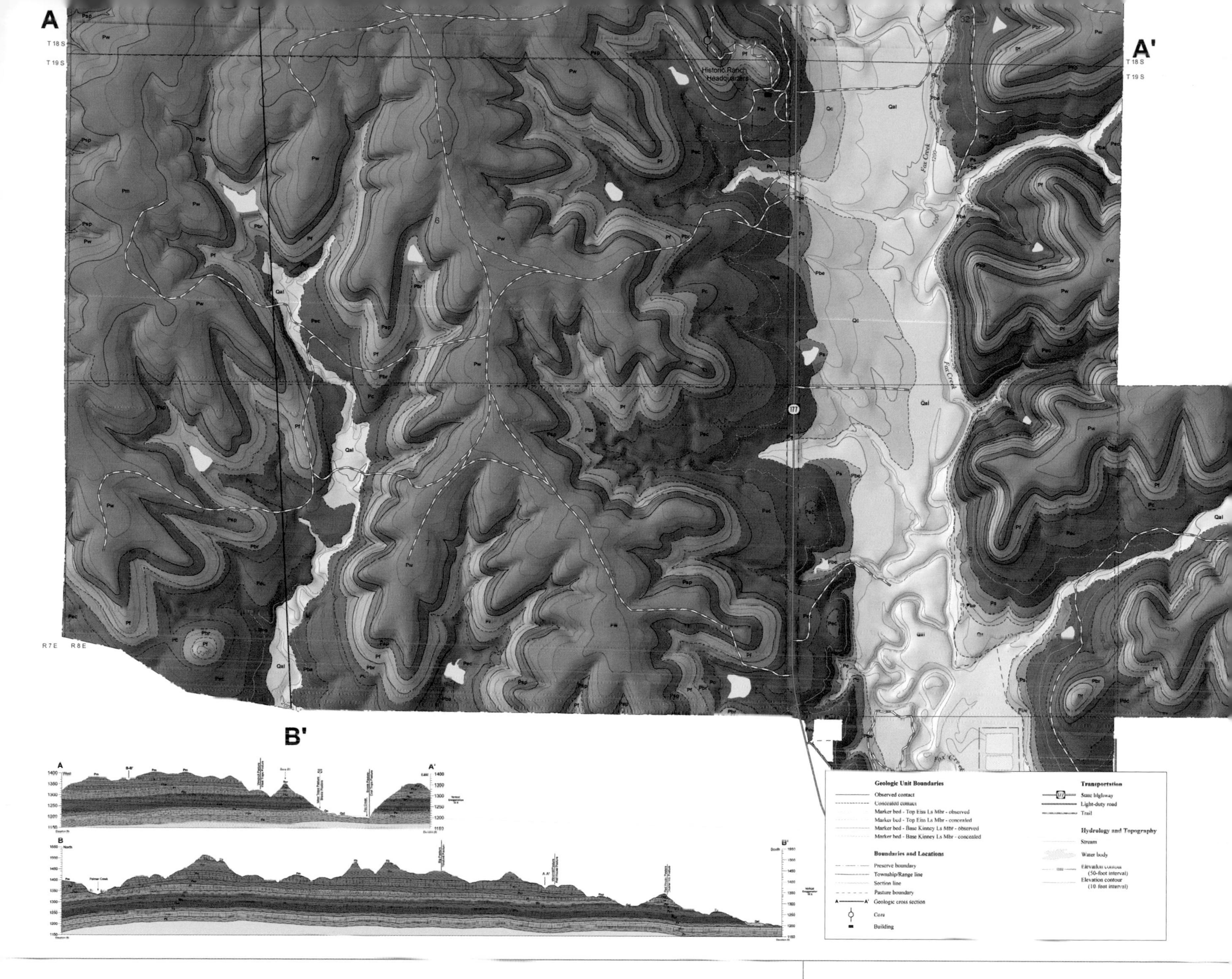

A better understanding of the geology and water resources of the Tallgrass Prairie National Preserve, located in the Flint Hills physiographic region of Kansas, is critical to long-term management of the preserve's natural resources. The newly created geologic map, hydrogeologic map, and supporting data provide the National Park Service and The Nature Conservancy (a public-private partnership jointly managing the preserve) with benchmark information about the preserve's geology and natural resources. Because the preserve's land-use patterns—including burning regimes, grazing, and human visitation—are changing, it is important to have baseline geologic and hydrogeologic data. The knowledge gained at the preserve will be useful to resource managers, researchers, and others at the preserve and throughout the Flint Hills region.

This surficial geology map shows the distribution, rock type, and age of bedrock near the earth's surface. It can be used to identify surface and subsurface lithologic units and their stratigraphic relationships, show geologic structures, delineate thick surficial materials such as alluvium, and show the spatial orientation of these features. The geologic map will be useful in construction and engineering projects, in understanding ground-water characteristics, and for environmental assessments. Understanding the near-surface geology and incorporating geologic evaluations into the planning process can help prevent future construction, resource, and environmental problems.

Courtesy of Kansas Geological Survey.

Kansas Geological Survey
Lawrence, Kansas, USA
By John Dunham, Robert Sawin, Nathaniel Haas, and Darren Haag

Contact
John Dunham
dunham@kgs.ku.edu

Software
ArcGIS Desktop 9.2, ArcGIS Spatial Analyst, Profile Tool 1.2.1

Printer
HP Designjet Z6100ps

Data Sources
Multiple sources, including Robert Sawin's field mapping

Surficial Geology of the Tallgrass Prairie National Preserve, Chase County, Kansas

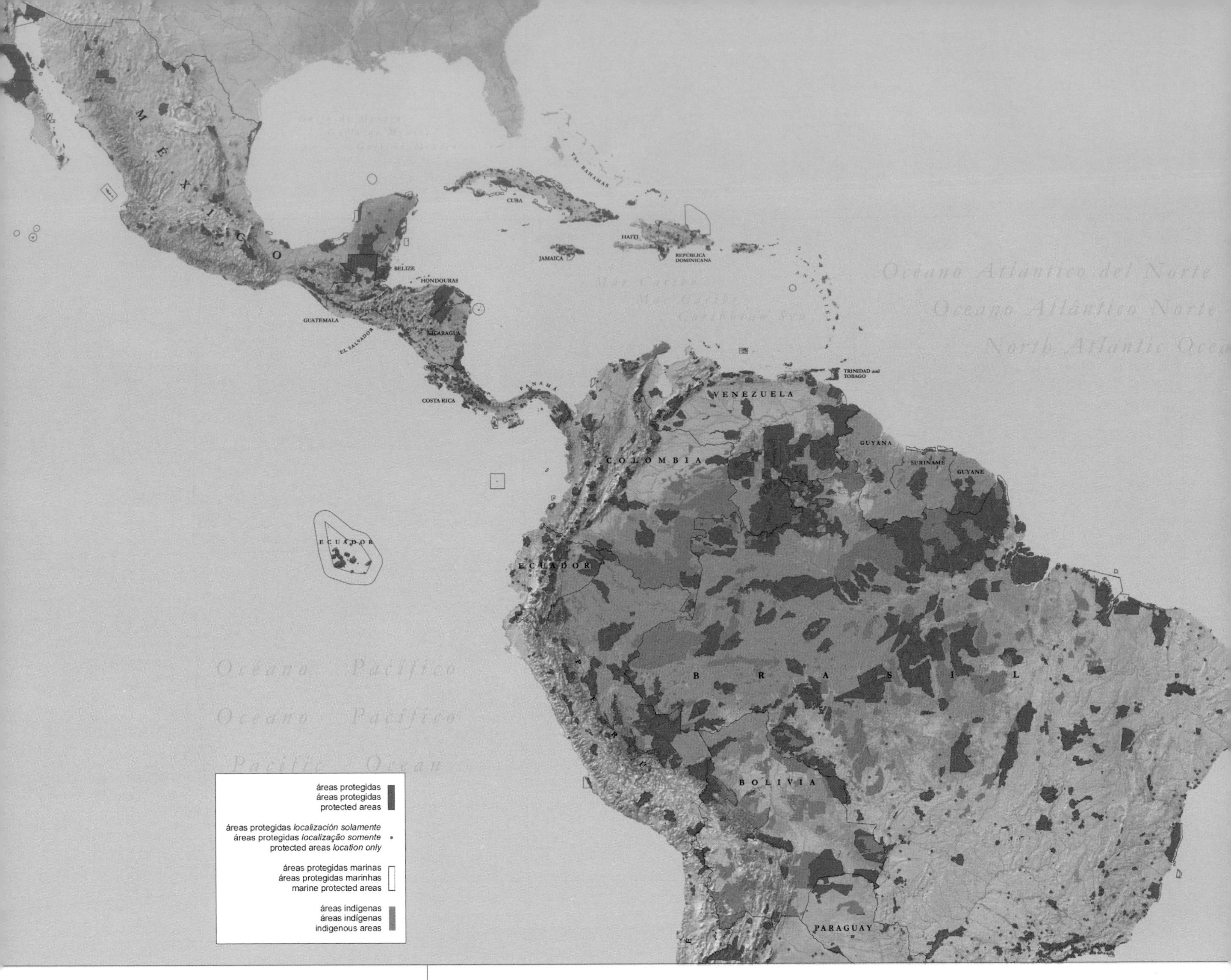

Conservation International
Arlington, Virginia, USA
By Mark Denil

Contact
Kellee Koenig
kkoenig@conservation.org

Software
ArcGIS Desktop 9.2

Printer
HP Designjet 5500ps

Data Sources
Conservation International, United Nations Environmental Programme—World Conservation Monitoring Centre, Shuttle Radar Topography Mission, ESRI

This map of neotropical protected areas, including indigenous areas, was prepared for the Latin American Congress of National Parks and Other Protected Areas, held in Bariloche, Argentina, September 30 to October 6, 2007.

Data for the protected and indigenous areas came largely from the United Nations Environment Programme—World Conservation Monitoring Centre's World Database of Protected Areas, with significant additions from sources in Conservation International. Elevation data from the National Aeronautics and Space Administration's Shuttle Radar Topography Mission. Drainage and political boundaries from ESRI.

Information on the map is presented in Spanish, Portuguese, and English, as was appropriate for this important international conference. The large-format (36 × 34-inch) map shows the neotropical protected areas of the western hemisphere at a scale of 1:9,700,000, and uses an equal area azimuthal projection centered on 77°W longitude and 5°S latitude. The key map, at a scale of 1:100,000,000, is in the same azimuthal projection.

Courtesy of Conservation International.

Neotropical Biodiversity Protected Areas: Version 2

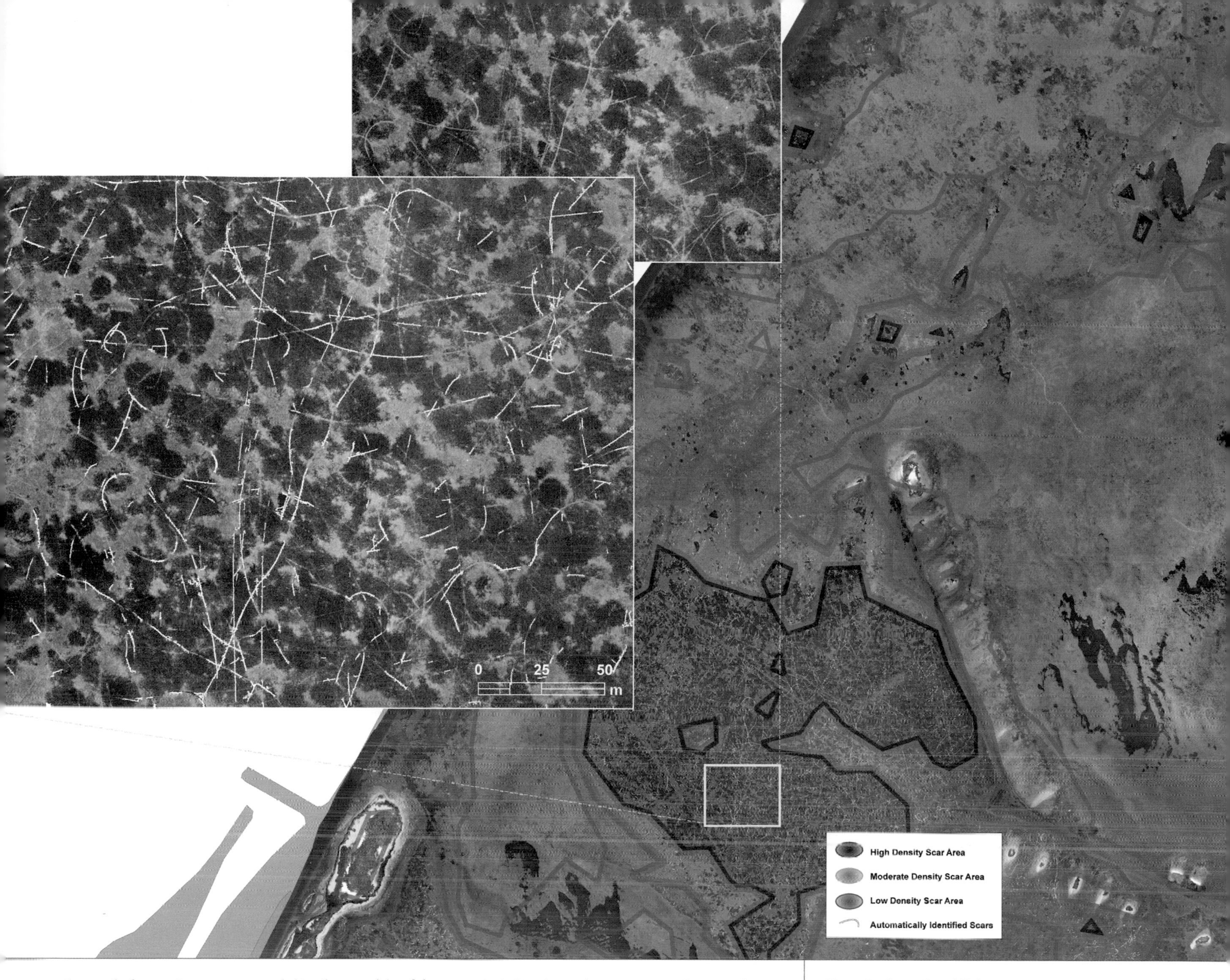

Seagrass beds are an important nursery habitat for many fish and shrimp species in Texas' coastal waters. Boaters often cause damage in these shallow areas by scarring seagrass beds with their propellers, leaving long "scars" or bare areas.

Due to the sensitive nature of this precious resource, the Texas Parks and Wildlife Department (TPWD) monitors seagrass health carefully. In 2005, the Texas Parks and Wildlife Commission passed a law prohibiting the uprooting of seagrass in Redfish Bay and TPWD Coastal Fisheries staff began an intensive study of seagrass scarring trends. To observe scarring behavior, TPWD acquired high-resolution imagery (0.1m) in 2007 intending to make comparisons with imagery in 2008 and 2009.

This map represents phase 1 of the assessment, in which automated feature detection software was employed to quickly identify scarred areas. A data mining tool was then used to remove commission errors from the Feature Analyst output. (This methodology was developed by Kass Green, Mark Tukman, and Mark Finkbeiner in the 2008 Redfish Bay Texas Airborne Sensor Comparison and Propeller Scar Mapping Final Report.) Parameters were changed to suit the needs of this project.

Using the ArcGIS Spatial Analyst extension, kernal densities were created from the centroids of the scar polygons. The resulting map identifies areas of high-, moderate-, and low-scarring intensities.

Texas Parks and Wildlife Department
Austin, Texas, USA
By Ashley Summers

Contact
Ashley Summers
Ashley.summers@tpwd.state.tx.us

Software
ArcGIS Desktop, ArcGIS Spatial Analyst, VLS Overwatch Geospatial, VLS Feature Analyst for ArcGIS, Rulequest See5/C5.0

Printer
HP Designjet 5000 ps

Data Source
Texas Parks and Wildlife Department

Automated Scar Extraction and Kernal Density Analysis

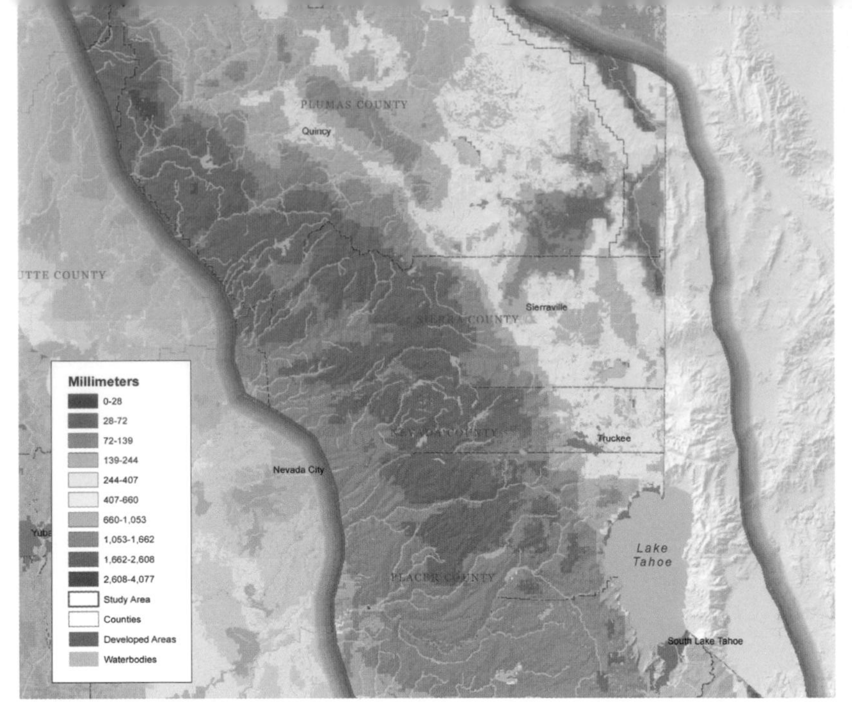

Average Annual Water Yield

Average Annual Water Retention

Carbon Storage (Live Tree Biomass)

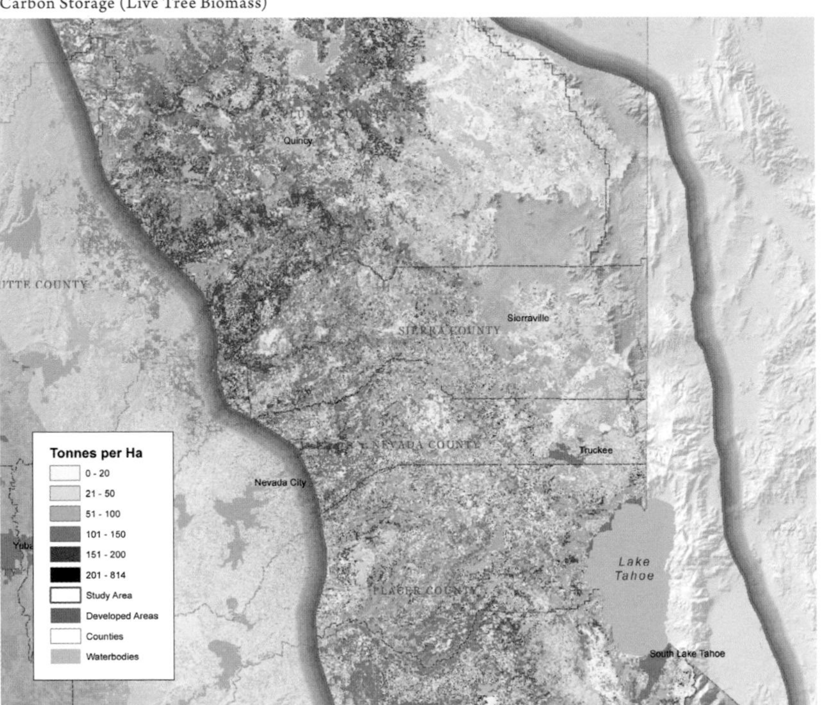

Biodiversity: Habitat Quality

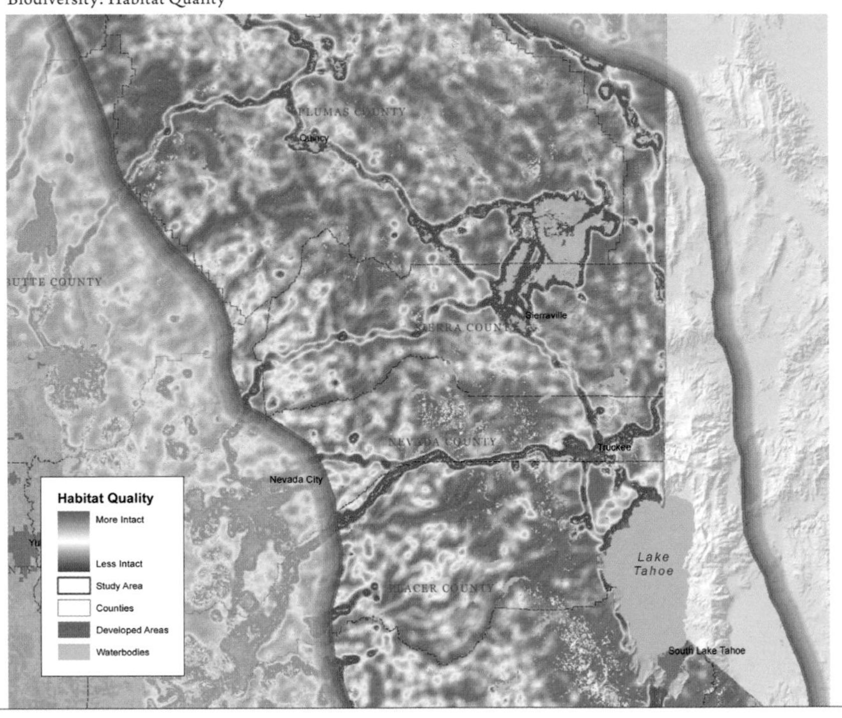

The Nature Conservancy
San Francisco, California, USA
By Erik Haunreiter and Dick Cameron

Contact
Erik Haunreiter
ehaunreiter@tnc.org

Software
ArcGIS Desktop 9.2

Printer
HP Designjet 800ps

Data Sources
State Soil Geographic Database Soils Datasets; Environmental Protection Agency EMAP Polygon Layers; Oregon State University PRISM Grids; U.S. Department of Agriculture Forest Service California Vegetation Maps

Natural ecosystems provide many services ranging from necessities, such as food and water, to services, such as erosion control, flood regulation and storm protection, to cultural values of open space for recreation and spiritual renewal. The Natural Capital Project, a partnership between Stanford University, The Nature Conservancy, and the World Wildlife Fund, is developing, testing, and applying innovative methods for mapping ecosystem services, including water, agricultural production, carbon sequestration, and pollination.

The northern Sierra Nevada, which extends from south of Lake Tahoe to Lassen Volcanic National Park, is home to exceptional natural, cultural, and recreational resources of statewide and global significance. The region also faces immediate threats from development and catastrophic wildfire as well as the likelihood that global climate change will significantly affect the region's natural resources.

Shown here are four ecosystem services in the Northern Sierra Nevada: water yield, water retention, carbon sequestration, and biodiversity. These maps represent the first step in mapping the delivery, distribution, and economic value of ecosystem services in the Sierra Nevada of California.

Mapping Ecosystem Services in the Sierra Nevada, California

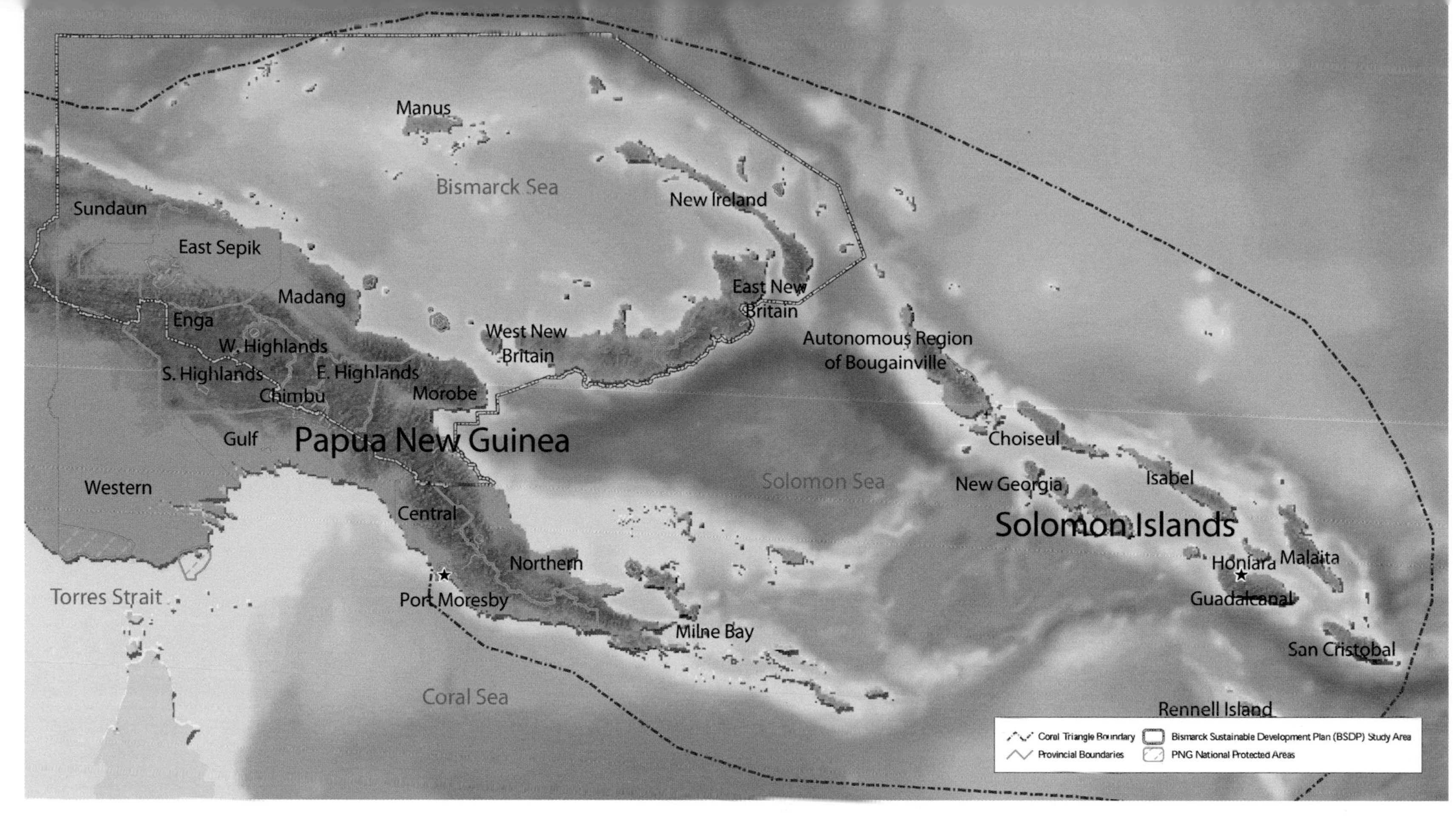

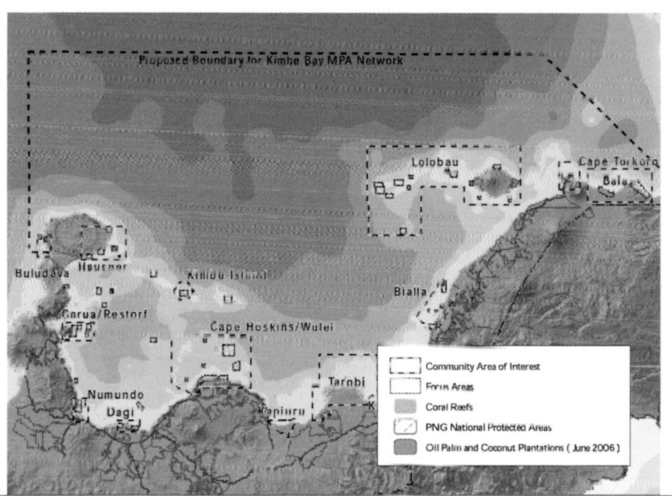

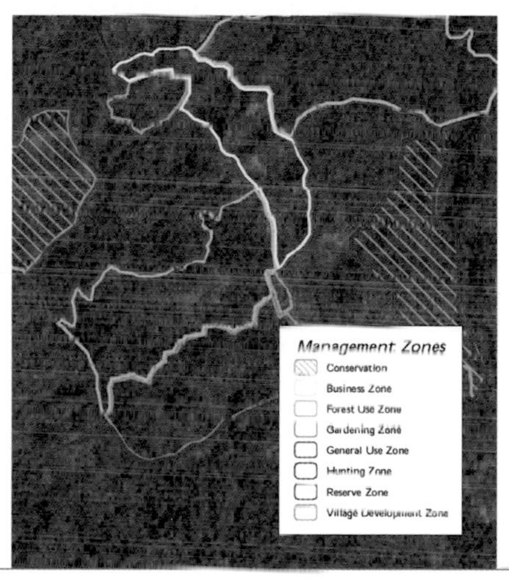

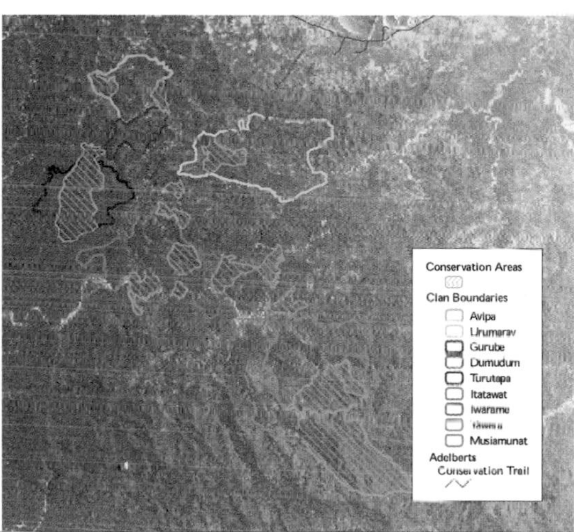

Papua New Guinea and the Solomon Islands are home to magnificent forests and coral reefs. This map highlights the rich habitats with a focus on current conservation action. Marine areas and terrestrial conservation areas are managed by the local people and government with support from The Nature Conservancy. The map also illustrates the multiple scales that The Nature Conservancy works in, from empowering local villages to coordinating across the region with the Coral Triangle.

Courtesy of Nate Peterson and the Melanesia Team, The Nature Conservancy.

The Nature Conservancy
South Brisbane, Queensland, Australia
By Nate Peterson

Contact
Nate Peterson
npeterson@tnc.org

Software
ArcGIS Desktop 9.2

Printer
HP Designjet 500

Data Sources
Shuttle Radar Topography Mission (SRTM) digital elevation model, Papua New Guinea Government Offices, The Nature Conservancy program data

Conservation Action in Melanesia

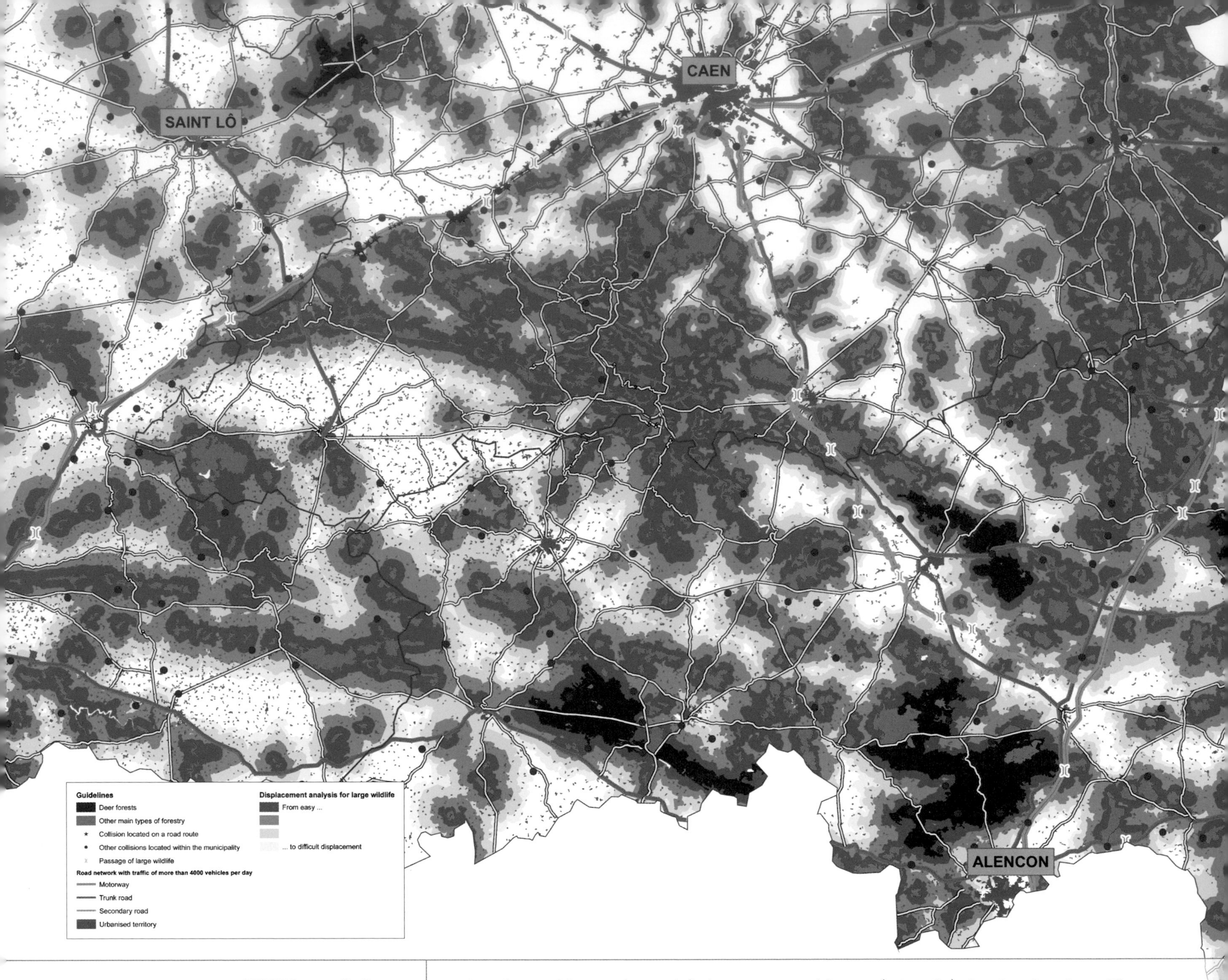

CETE Normandie Centre
Grand Quevilly, France
By Pierre Vigne and Jean Francois Bretaud

Contact
Pierre Vigne
pierre.vigne@developpement-durable.gouv.fr

Software
ArcGIS Desktop, GRID, TIN

Printer
HP Designjet 1050

Data Sources
BD Topo, Corine Land Cover

With nearly 10,000 kilometers (6,214 miles) of motorway, 25,500 kilometers (15,845 miles) of trunk roads, and 1,600 kilometers (994 miles) of high-speed train lines in Basse-Normandie, transport networks dividing up natural habitats can have two main effects on animal and plant species. One, they reduce the size of the habitat such that populations of species with large home ranges can no longer survive in them. The other effect is the isolation of the remaining habitat patches, such that species have little chance of moving from one to the other.

In this situation, the species concerned are threatened with local or regional extinction. It is through these processes that habitat fragmentation by transport networks and the resulting secondary phenomena have become the most serious threats to biological diversity on the planetary scale.

As part of the French national strategy for biodiversity, and in response to the alarming report on the assessment of the A84 motorway concerning collisions with wildlife, the Environment and Geomatics department of the CETE Normandie Centre set up this study of ecological networks in order to propose development plans in favor of the species concerned and to improve the safety of road users.

Basse-Normandie Seen through a Deer's Eyes

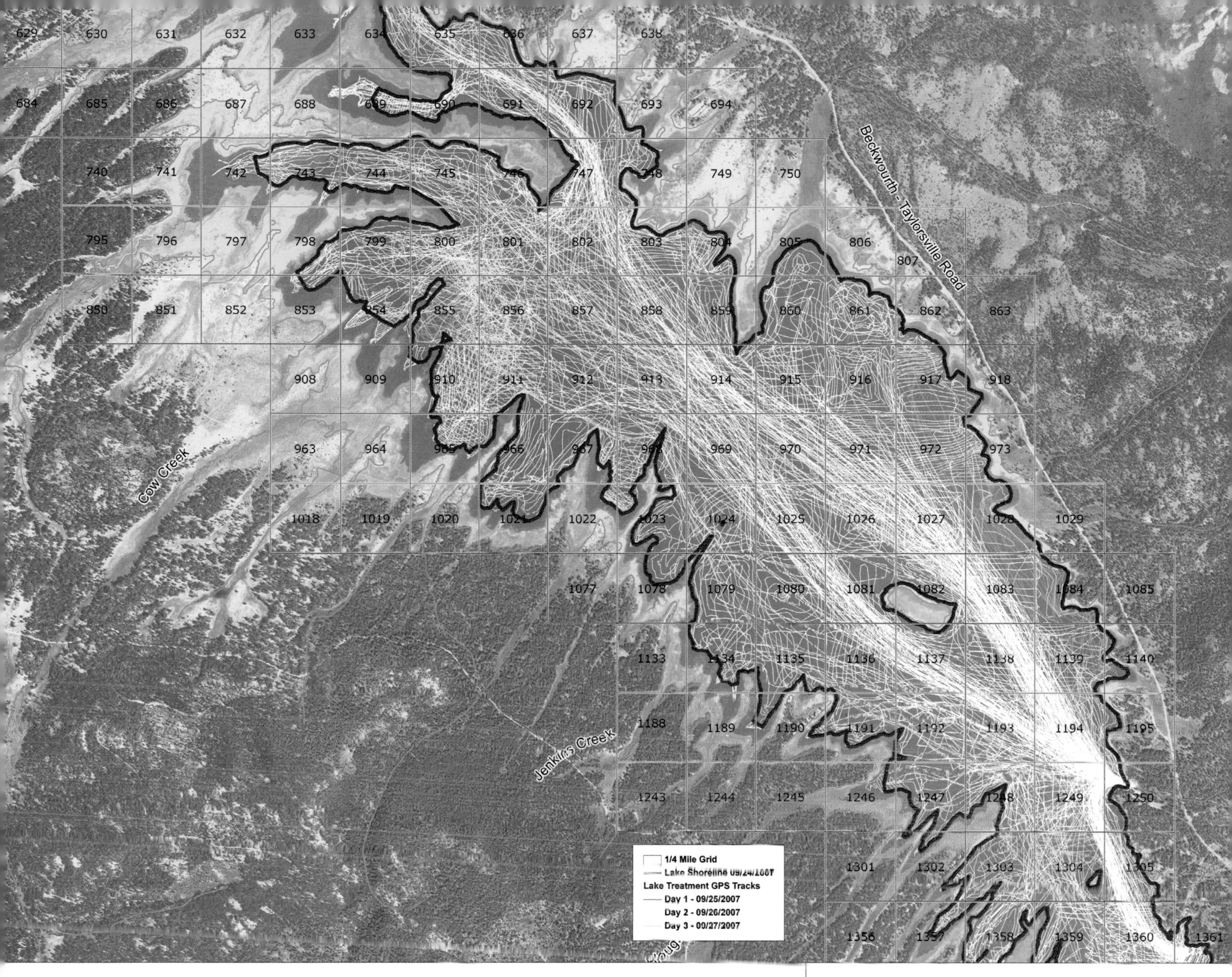

Global Positioning System tracks on Lake Davis represent how the California Department of Fish and Game applied the chemical rotenone to rid the reservoir of the predatory and invasive northern pike *(Esox lucius)*. Pike have adversely affected the Lake Davis trout fishery and the associated local economy. If pike escape or are moved from the reservoir, they could endanger fish populations in other waters where they become established. For example, pike have the potential to cause irreversible damage in portions of California's Feather, Sacramento, and San Joaquin river systems; the Sacramento-San Joaquin Delta; as well as many other waters of the state and region.

Courtesy of California Department of Fish and Game.

California Department of Fish and Game, South Coast Region
San Diego, California, USA
By Ken DeVore

Contact
Ken DeVore
kdevore@dfg.ca.gov

Software
ArcGIS Desktop 9.2

Printer
HP Designjet 1055cm

Data Sources
Garmin database files

2007 Lake Davis Pike Eradication Project—Lake Treatment

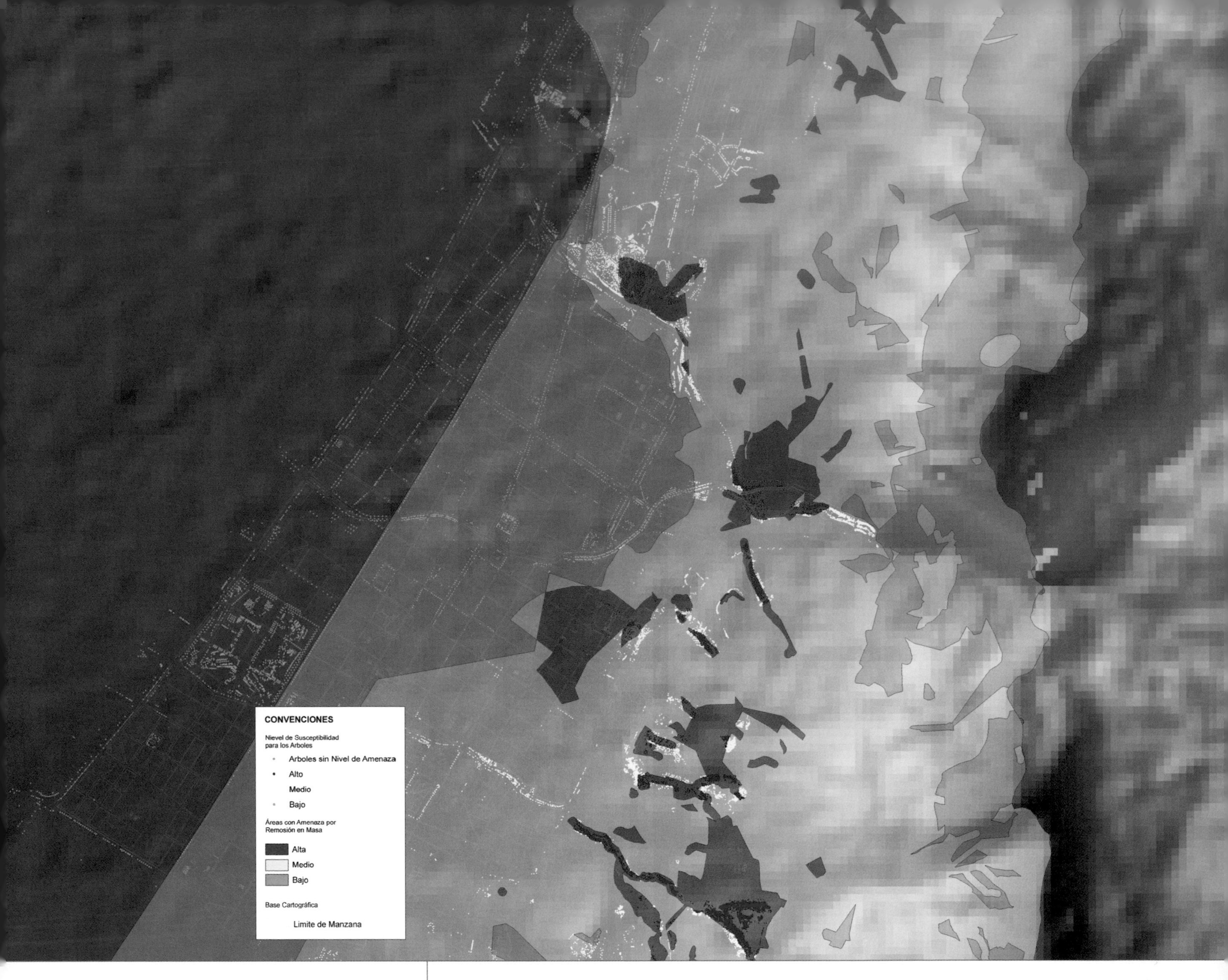

Alcaldia Mayor de Bogotá
Bogotá, D.C., Bogotá, Columbia
By Amy Uribe Navarro

Contact
Amy Uribe Navarro
amyuribe@gmail.com

Software
ArcGIS Desktop 9.2

Printer
HP DesignJet 1055 cm

Data Sources
Corporation geodatabase, 30-by-30m DEM of Bogotá.

This map shows the risk for mass removal and the susceptibility level for trees. It contains city blocks and shadow relief created on the Bogotá digital elevation model (DEM). Mass removal occurs when large quantities of ground slide and shift after earthquakes, tremors, or heavy rains. Mass removal seriously damages urban infrastructures and, in many cases, threatens human life.

The map indicates trees located in zones of risk for mass removal as well as the relief of the zone. The map is a valuable tool for the entities responsible for preventing and responding to mass removal. It shows the places where they have to intervene.

Courtesy of Jardín Botánico José Celestino Mutis.

Analysis of Susceptibility for Threat (Removal) in Mass in the Locality of Santa Fe, Bogotá

This map analyzes biological corridors (linear strips of vegetation that provide a continuous or near-continuous pathway between habitats) and tree crowns (the area above the trunk) in the locality of Chapinero, Bogotá. It shows the trees modeled with an equatorial diameter buffer and integrated with a dissolve for biological research. The trees with their crowns close together serve as a biological corridor for local birds and insects. They also provide shade for people living in the city.

Courtesy of Jardín Botánico José Celestino Mutis.

Alcaldia Mayor de Bogotá
Bogotá, D.C., Bogotá, Colombia
By German Herreño Fierro

Contact
German Herreño Fierro
gherreno@gmail.com

Software
ArcGIS Desktop 9.2

Printer
HP Designjet 1055cm

Data Source
Censo del Arbolado Urbano de Bogotá

Alliance for a Better Georgina
Keswick, Ontario, Canada
By Jason Anderson, Mark Setter, Daniel Faucher

Contact
Jason Anderson
jason@georginamaps.ca

Software
ArcGIS Desktop 9.1, Adobe Illustrator, MAPublisher, Adobe Photoshop, Geographic Imager

Data Sources
Geobase, Alliance for a Better Georgina, York Region, Lake Simcoe Region Conservation Authority

This map is part of a series of community maps of the town of Georgina produced by residents. Its purpose is to showcase what residents think is important about their neighborhoods and to help protect and enhance what is special. The map points out a nature reserve and a local stream, both important habitats for conservation.

The Deer Park Road area is an important mature mixed forest providing habitat for a wide range of animals, birds, and amphibians. Private landowners in the area worked together with the Lake Simcoe Region Conservation Authority to create the Arnold C. Matthews Nature Reserve in order to preserve and steward the land in perpetuity.

Boyer's Stream connects protected core lands and Lake Simcoe, providing habitat for waterfowl, amphibians, and marsh birds, and a potential upstream fish spawning route.

Courtesy of the Alliance for a Better Georgina.

For the Love of the Lake, Historic Lakeshore Communities

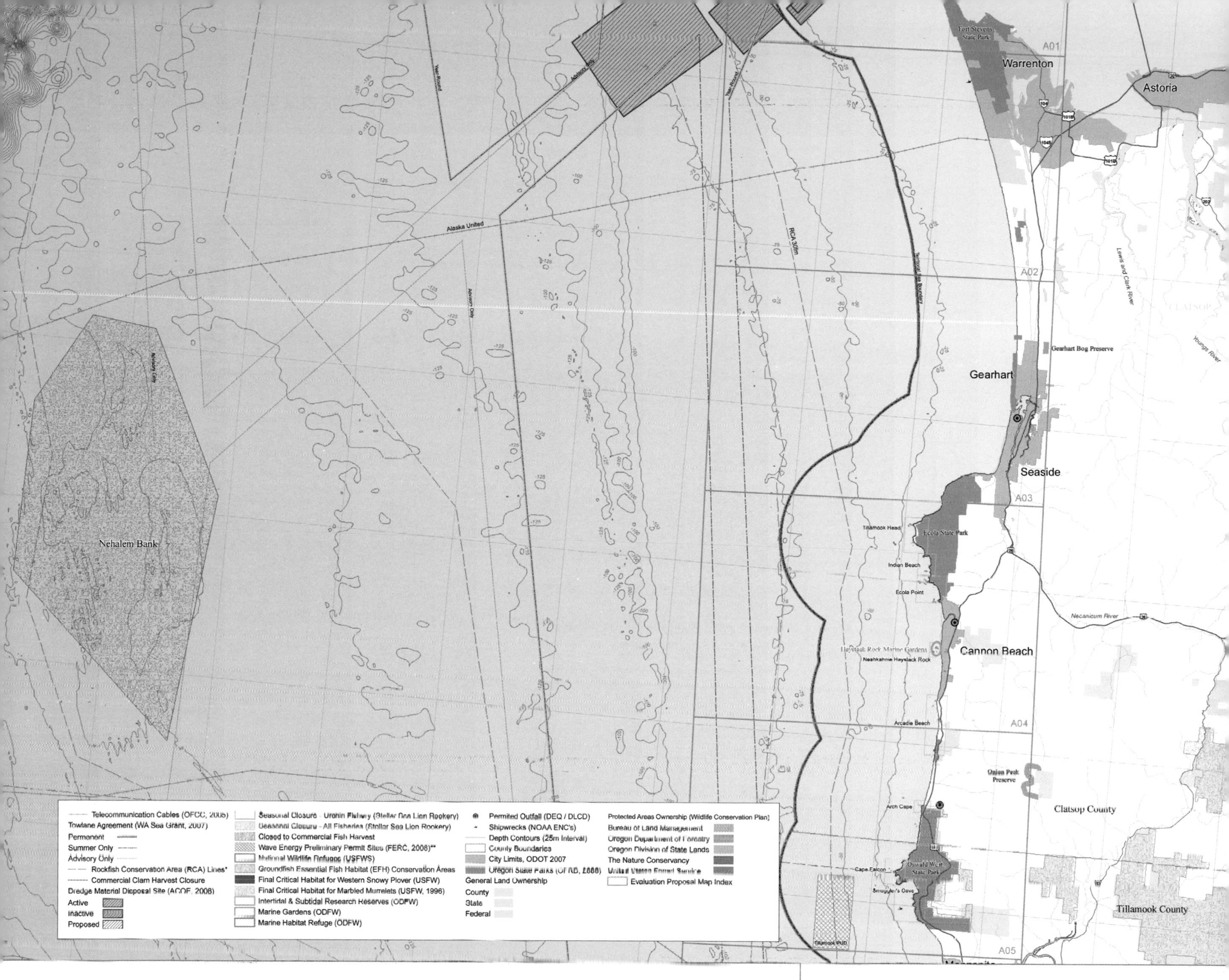

The State of Oregon and Oregon's Ocean Policy Advisory Council (OPAC) need geospatial information for the coastal and offshore areas of Oregon for planning. This map is one in a reference chart series showing human use and management. Generated initially to support the Oregon marine reserve proposal process, these maps were part of a larger effort that also mapped biological resources and seafloor and shoreline.

The sea has traditionally been familiar territory to the fishing community. Other interests such as potential wave energy projects and marine conservation have emerged. The maps help OPAC members and others understand current and potential use of specific areas. The maps are used to facilitate planning, state and local discussions, and education efforts.

Courtesy of Andy Lanier and Barbara Seekins.

National Oceanic and Atmospheric Administration Fisheries, Northwest Region, Oregon Department of Land Conservation & Development

Portland, Oregon, USA

By Barbara Seekins and Andy Lanier

Contact

Barbara Seekins, barbara.seekins@noaa.gov

Software

ArcGIS Desktop, MapBook

Printer

HP Designjet 1055 cm Plus, 36 inch

Data Sources

National Oceanic and Atmospheric Administration, Oregon Department of Land Conservation and Development, Oregon Department of Fish and Wildlife, U.S. Army Corps of Engineers, Oregon State University, U.S. Fish and Wildlife Service, Oregon Fisherman's Cable Committee

Human Use and Management Chart, North Coast Section of Oregon

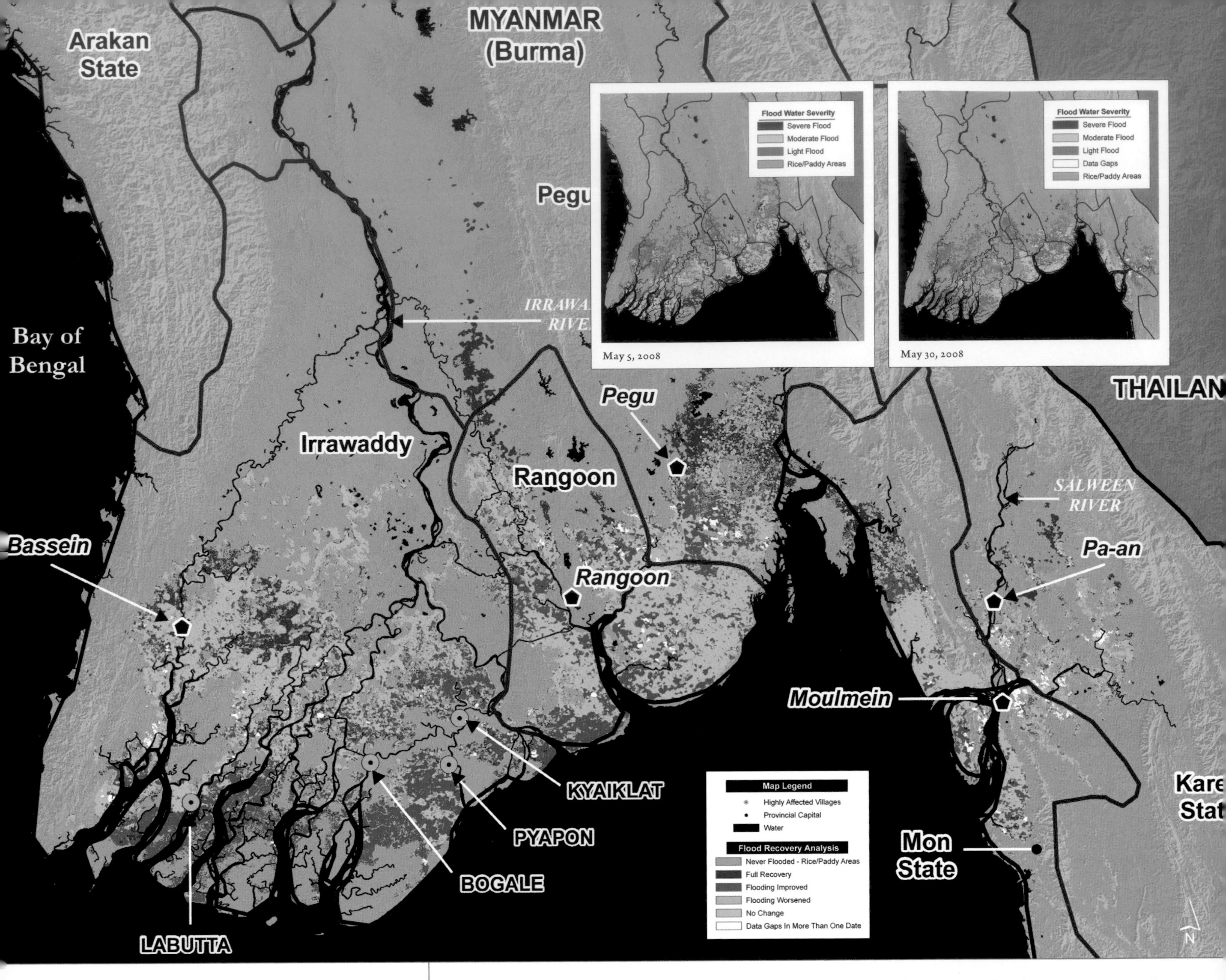

U.S. Department of Agriculture, Foreign Agricultural Service
Washington, D.C., USA
By Marlise Wilson, Michael Shean, and Sean Griffin

Contact
Marlise Wilson
Marlise.Wilson@asrcms.com

Software
ArcGIS Desktop 9.2

Printer
HP Designjet 5500ps

Data Sources
U.S. Department of Agriculture, ASRC Management Services, Geospatial Data Analysis Corporation, ESRI, National Aeronautics and Space Administration, U.S. Geological Survey Earth Resources and Observation Science Center, Vector Map, GeoCover LC 2000

Tropical cyclone Nargis struck the heart of Burma's rice growing region in the low-lying Ayeyarwady Delta on May 2, 2008, causing extensive damage to agricultural lands, infrastructure, livestock, and stored food grains. A nearly 2,000-square-mile area of prime farmland was inundated with salt water and/or heavy rainfall. The affected region normally accounts for roughly 60 percent of the nation's rice production.

The U.S. Department of Agriculture (USDA) had conducted a post-flood assessment that indicated that as of May 30, 2008, flood waters receded over a sizable area (300,000 hectares total recovery; 490,000 hectares improved since May 5, 2008). However, a month after the cyclone, approximately 1.40 million hectares, or 80 percent of the original inundated area, was still affected by some degree of flooding. Approximately 870,000 hectares had shown no improvement. The areas that showed the greatest change in the severity of flooding were the coastal areas of southern Ayeyarwady division. Natural drainage in these coastal rice farming areas must have aided the recovery, as further inland crop areas did not show the same degree of improvement. In contrast, much of the southern regions of Yangon division, which were heavily inundated, did not show much improvement a month after the cyclone's passing.

The Foreign Agricultural Service of the USDA works to improve foreign market access for U.S. products, build new markets, improve the competitive position of U.S. agriculture in the global marketplace, and provide food aid and technical assistance to foreign countries.

Myanmar (Burma) Cropland Recovery and Severity Analysis of Tropical Storm Nargis—2008

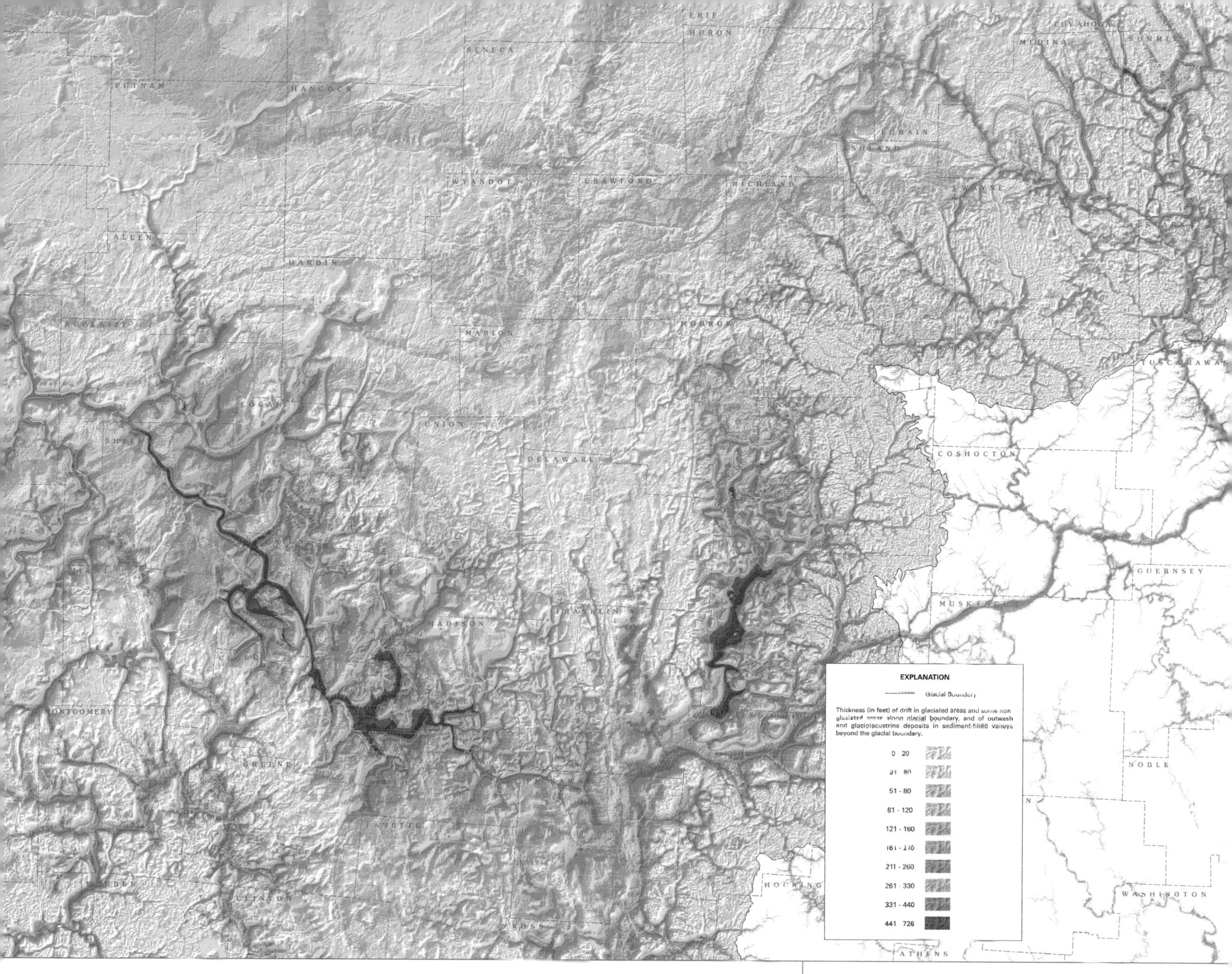

The drift-thickness map of Ohio depicts the thickness and distribution of glacially derived sediments (called drift) and post-glacial stream sediments overlying the buried bedrock surface. This map was produced by subtracting bedrock-surface elevations from land-surface elevations to produce a residual map of drift thickness. Colors portray thickness intervals of glacial and modern sediments, which can range up to several hundred feet.

The bedrock-surface component is one of the products resulting from a multiyear effort by the Ohio Department of Natural Resources, Division of Geological Survey, to map the bedrock geology of Ohio. Bedrock-topography maps are required to determine the relief on the bedrock surface beneath thick layers of glacial drift. These maps were created for all 788 7½-minute topographic quadrangles in the state as part of a process to produce accurate bedrock-geology maps for glaciated portions of Ohio and for those areas beyond the glacial boundary where valleys are infilled with sediment. Data concentration and contour intervals on the original, hand-drawn bedrock-topography maps vary widely across the state in response to changing geologic and topographic conditions. During the course of mapping, over 162,000 data points were interpreted for bedrock-surface elevation and in some cases drift thickness. These points were plotted on maps and used as control for the bedrock-topography lines.

Courtesy of Donovan Powers, Ohio Department of Natural Resources, Division of Geological Survey.

State of Ohio

Columbus, Ohio, USA
By Donovan M. Powers and E. Mac Swinford

Contact

Donovan M. Powers
donovan.powers@dnr.state.oh.us

Software

ArcGIS Desktop, Adobe Photoshop, and Adobe Illustrator

Printer

HP Designjet 5500ps (ultra violet ink)

Data Sources

Ohio Department of Natural Resources (Division of Geological Survey, Division of Water, Division of Mineral Resources Management), Ohio Department of Transportation, Ohio County Engineers, and U.S. Geological Survey

Shaded Drift-Thickness Map of Ohio

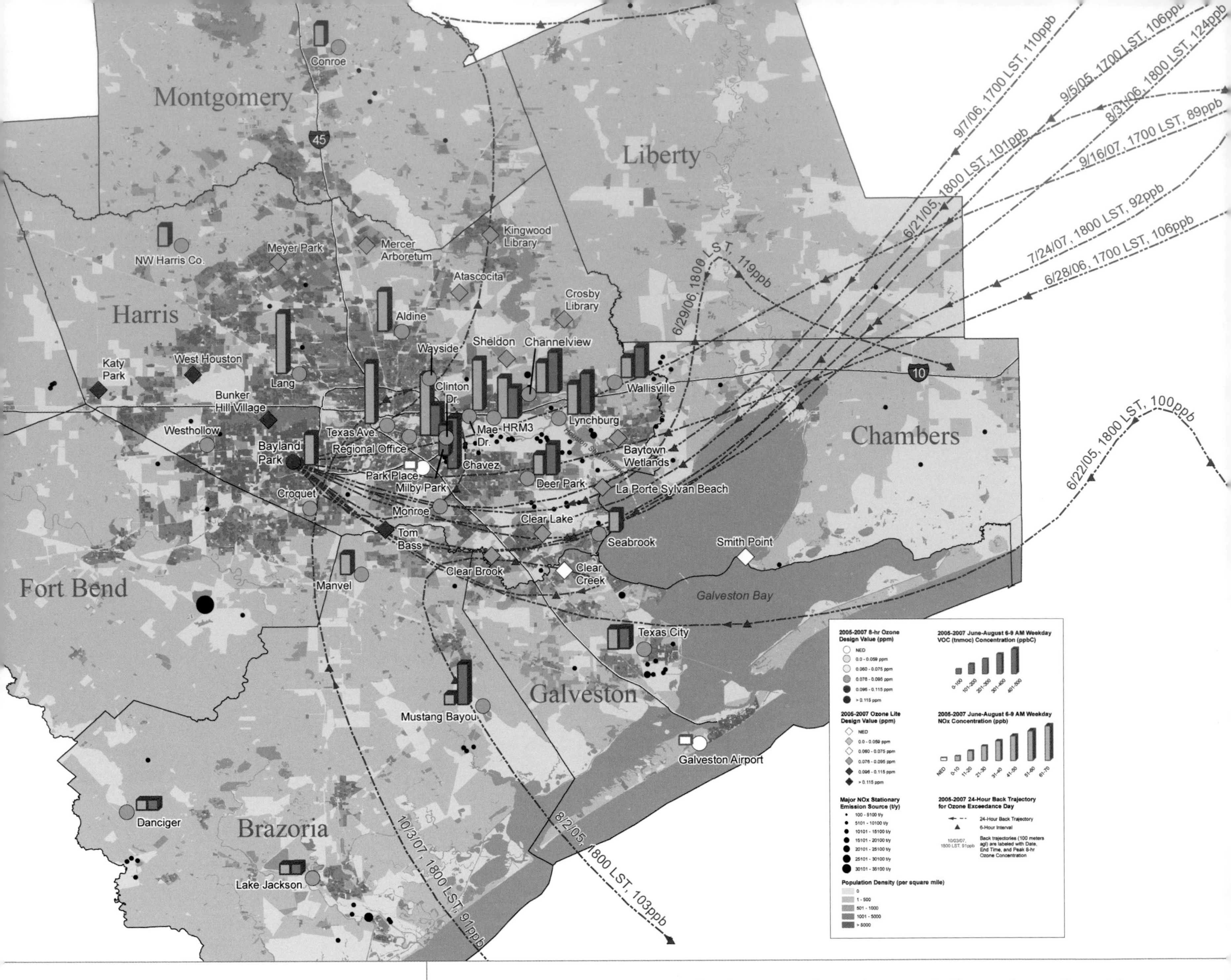

U.S. Environmental Protection Agency Region 6
Dallas, Texas, USA
By Melody Lister (Lockheed Martin) and Mark Sather (EPA)

Contact
Mark Sather
sather.mark@epa.gov

Software
ArcGIS Desktop 9.2

Printer
HP Designjet 5500ps

Data Sources
U.S. Environmental Protection Agency, U.S. Census Bureau, National Oceanic and Atmospheric Administration

The U.S. Environmental Protection Agency (EPA) Region 6 needs to analyze regional pollution to better understand the ozone problem in key areas. This map assists in the study by showing the spatial relationship between major nitrogen oxides (NOx) stationary emission sources, ambient ozone concentrations, ambient NOx and volatile organic compounds (VOC) precursor concentrations, population density, and 24-hour wind back trajectories on high ozone concentration days.

The back trajectories, created using the National Oceanic and Atmospheric Administration's Hysplit Model, help identify possible source areas of the contaminant. This enables EPA Region 6 to target areas for ozone reduction more efficiently.

As shown here, none of the monitors in the Houston area met the 2008 8-hour ozone standard of 0.075 ppm. Ozone design values range from 0.076 ppm at Lake Jackson in southwestern Brazoria County to 0.096 ppm at Bayland Park west of downtown and the Ship Channel. The higher ozone concentrations in the western part of Houston are supported by the predominance of east sector wind flow on the highest ozone concentration days.

Courtesy of U.S. Environmental Protection Agency Region 6.

Houston Ozone and Ozone Precursor Monitoring Network

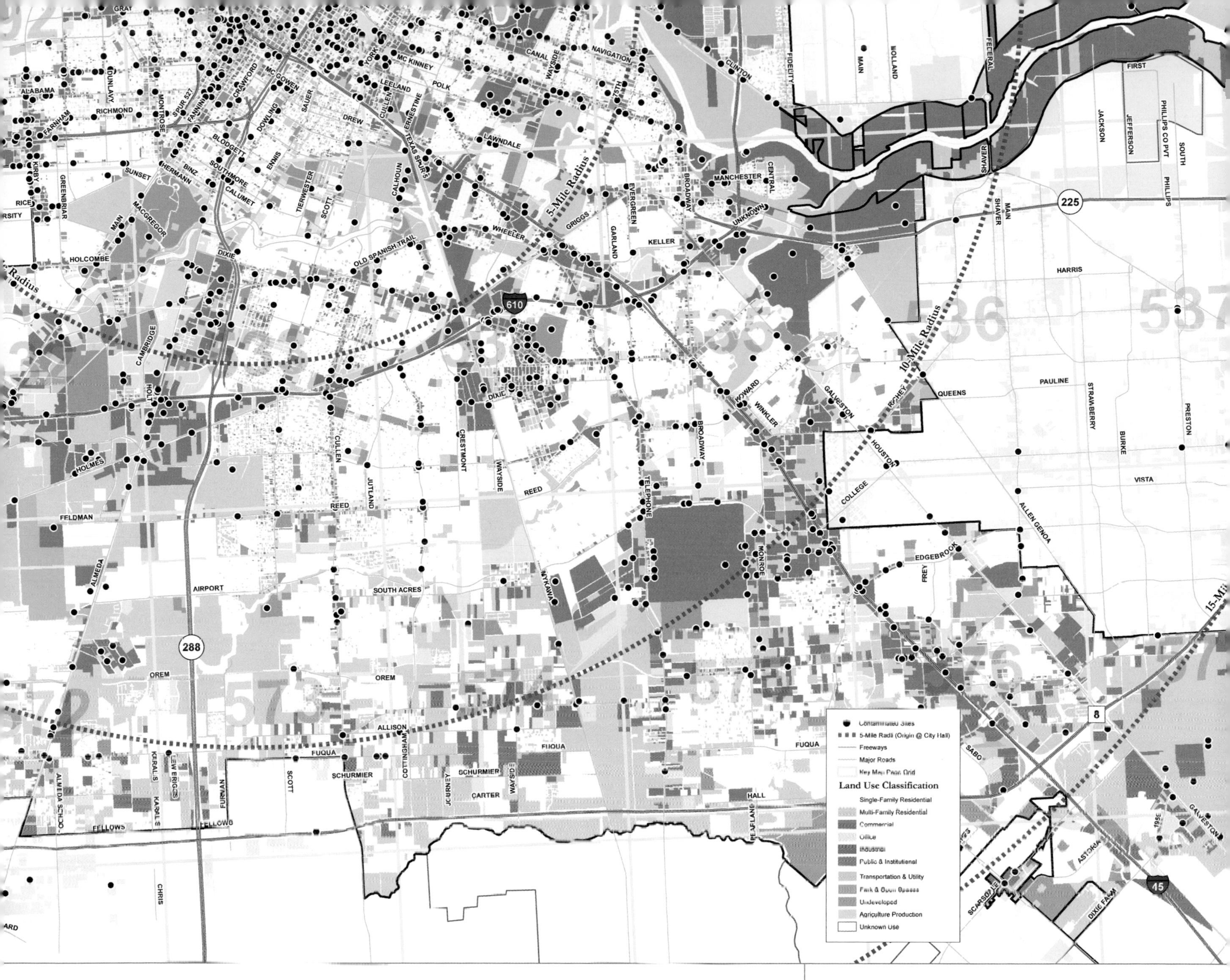

The City of Houston needed to locate all contaminated sites within its municipal boundary. Several agencies at the federal, state, and municipal levels tracked these locations, but there was no single source to show the entire universe of contaminated sites in the Houston area across all the programs. A comprehensive geodatabase would enable the city to analyze concentrations of these sites and prioritize locations to remediate.

Aggregating and analyzing these contaminated sites based on ZIP Codes, neighborhoods, council districts, and other boundaries gave policy makers vital data to better serve citizens. Houston's office of the mayor compiled the information from various environmental agencies and turned it over to the Planning Department's GIS mapping team. The mapping team took the data, which was broken down by participating program, and geocoded the addresses. Afterward, the feature classes were organized into a geodatabase by program affiliation and the concentrations of these sites were analyzed based on various regional boundaries. The final presentation map set summarizes the findings of the study, highlighting the concentrations of these contamination program sites based on known Houston geographical areas.

Courtesy of City of Houston Department of Planning and Development.

City of Houston Department of Planning and Development
Houston, Texas, USA
By Larry T. Nierth, GISP

Contact
Larry T. Nierth
larry.nierth@cityofhouston.net

Software
ArcGIS Desktop 9.2, ArcGIS 3D Analyst, Microsoft Access, Microsoft Excel, Adobe Photoshop

Printer
HP Designjet 1055cm

Data Sources
Texas Commission on Environmental Quality, Environmental Protection Agency, City of Houston

Contaminated Sites Geodatabase

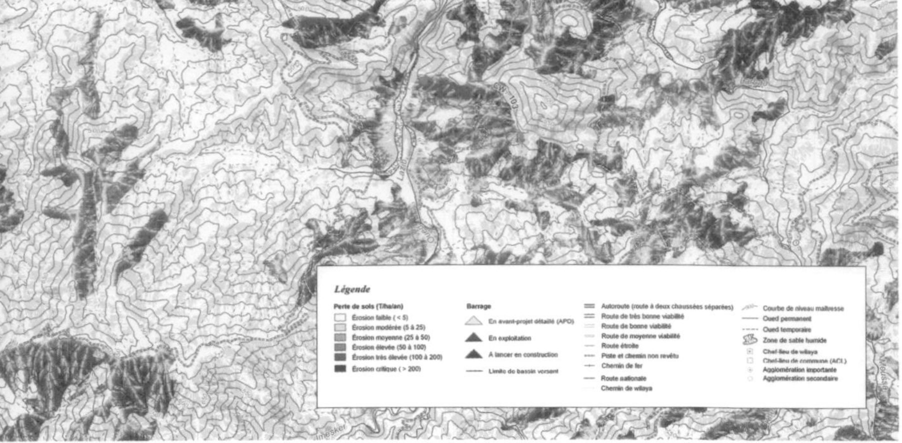

Sheet Erosion

Gully Erosion

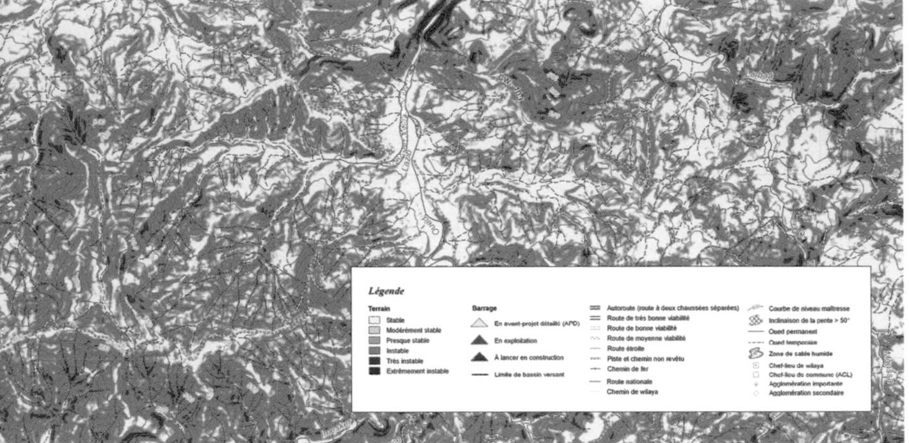

Landslides

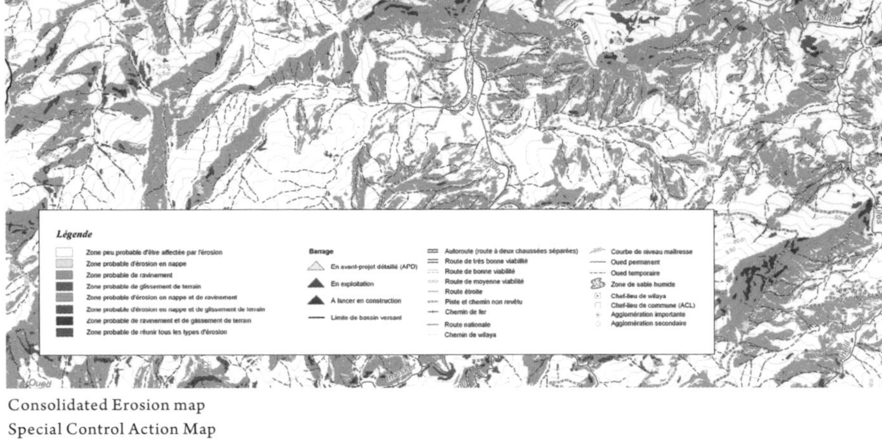

Consolidated Erosion map

Watershed Management Plan

Special Control Action Map

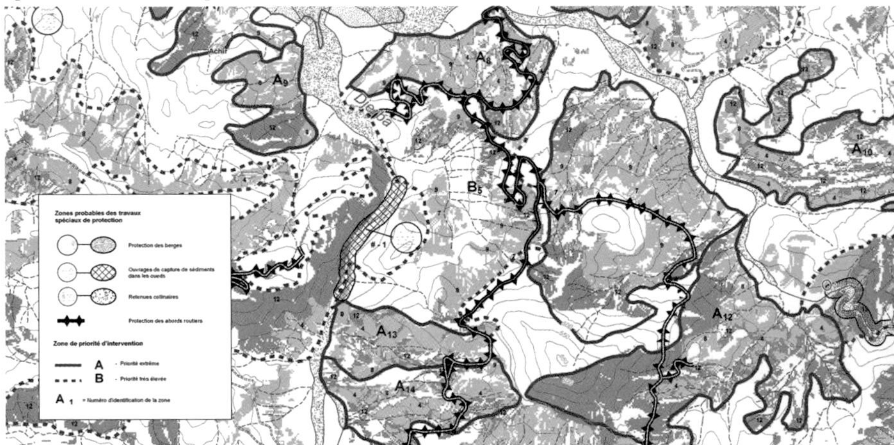

Tecsult Inc., AECOM
Québec City, Québec, Canada
By Guy Parent, François Trudeau, Sylvie Roy, André Lauzon, Pierre Roy, Henri Tichoux, Jacques Langlois, Martin Harvey, Salah Rechoum, Raphaël Fauchère, François Légaré, and Maria Fernanda Senia

Contact
Denis Baron
denis.baron@tecsult.aecom.com

Software
ArcGIS Desktop 9.2, ArcGIS Workstation, ArcGIS Spatial Analyst

Printer
HP Designjet T1100ps

Data Sources
Various

In Algeria, water is a key component of economic development, and its scarcity necessitates dams for storage and distribution for irrigation and human consumption. The Agence Nationale des Barrages et Transferts (ANBT) plans to build new dams to bring the total to 70 by 2010. However, soil erosion has contributed to a 20 percent reduction in reservoir capacities since their construction.

In 2003, the ANBT invited Tecsult Inc., a major Canadian engineering firm, to conduct a comprehensive study that would locate the degraded areas of watersheds of twenty-one dams (five existing and sixteen in the detailed planned stage) and to develop watershed management plans to reduce hillside erosion and decrease reservoir siltation (accumulation of silt). The total study area covered more than 23,800 square kilometers (9,190 square miles), and the allowed study time was eighteen months.

Each watershed was characterized for three types of erosion (sheet erosion, gully erosion, and landslides), and the resulting maps were combined in order to create a consolidated erosion risk map. Using this map and twelve land management measures, a watershed management plan was developed for each dam. Each watershed has also a map locating four kinds of special erosion control actions that aimed to reduce the stream sediment loads before they reach the reservoir. Finally, the budget required for implementing the watershed management plan with the special erosion control actions was estimated for each watershed. It was estimated that the twenty-one management schemes elaborated during this study will yield a total estimated gain of water in their reservoir equivalent to the consumption of water for more than 9 million people in one year.

Courtesy of Agence Nationale des Barrages et Transferts (ANBT).

Development of an Erosion-Reduction Management Strategy for Watersheds and Reservoirs in Algeria

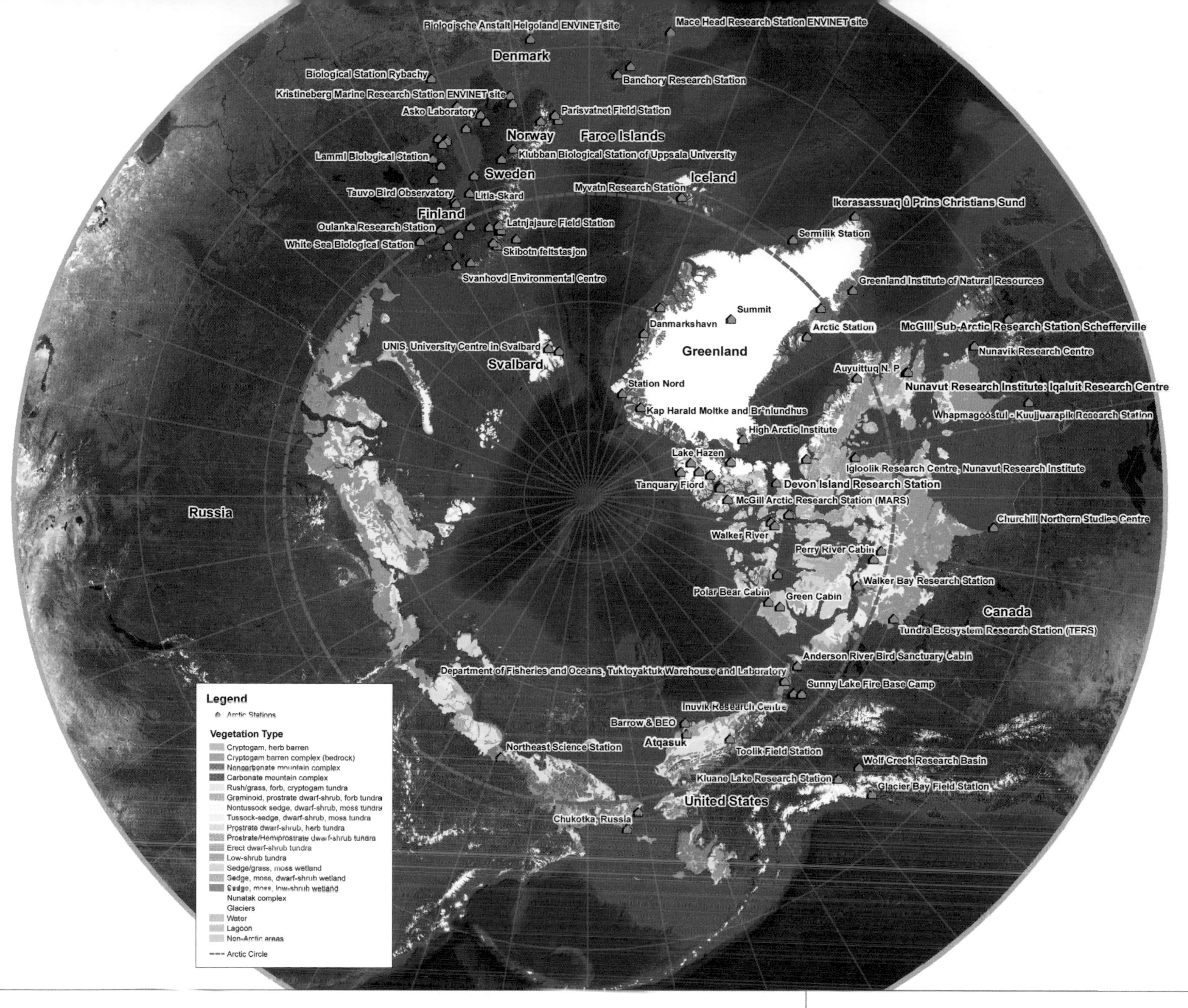

This map displays the Circumarctic Vegetation Map and Arctic Research Stations, which include the location, station name, contact information, facilities, equipment, communications, Internet link, and other basic information for stations supporting research in the Arctic region. The Arctic Station data layer is compiled by the Canadian International Polar Year Secretariat and was derived from the Canadian International Polar Year (CANIPY) Web portal and the terrestrial Circumarctic Environmental Observatories Network (CEON).

The map shows the types of vegetation that occur across the Arctic, between the ice-covered Arctic Ocean to the north and the northern limit of forests to the south. An international team of Arctic vegetation scientists representing the six countries of the Arctic (Canada, Greenland, Iceland, Norway, Russia, and the United States) prepared the map by grouping over 400 described plant communities into sixteen physiognomic units based on plant growth forms.

Both data layers can be visualized in the Arctic Research Mapping Application (ARMAP), a suite of online services, accessible at ARMAP.org, to provide support of Arctic science. With ARMAP's 2D maps and 3D globes, users can navigate to areas of interest, view a variety of map layers, and explore U.S. federally funded research projects. With special emphasis on the International Polar Year (IPY), ARMAP has targeted science planners, scientists, educators, and the general public.

Courtesy of ARMAP.org.

University of Texas at El Paso
El Paso, Texas, USA
By G. Walker Johnson

Contact
G. Walker Johnson
gjohnson@miners.utep.edu

Software
ArcGIS Desktop, ArcIMS 9.2 SP5

Printer
HP Designjet 4500

Data Sources
Arctic Research Logistics Support Service (ARLSS), Circumpolar Arctic Vegetation Map (CAVM), Arctic Stations (CANIPY/CEON), 500m Satellite NASA's Blue Marble Next Generation Images

Circumpolar Arctic Vegetation Map, Including Arctic Research Stations

Alaska Geobotany Center, Institute of Arctic Biology, University of Alaska, Fairbanks
Fairbanks, Alaska, USA
By Donald A. Walker and Hilmar A. Maier

Contact
Donald A. Walker
ffdaw@uaf.edu

Software
ArcGIS Desktop 9.2

Printer
Epson Stylus Pro 9800

Data Sources
Aerial photography (photo interpretation), SPOT imagery, geologic maps, photographic images

These vegetation maps are shown at three scales in the vicinity of the Toolik Field Station, Alaska, which is an Arctic research facility run by the Institute of Arctic Biology at the University of Alaska Fairbanks. The maps are intended to support research at the field station.

The front side of the map sheet contains a vegetation map and ancillary maps of a 751-square-kilometer (290-square-mile) region surrounding the upper Kuparuk River watershed, including the Toolik Lake and the Imnavait Creek research areas, as well as portions of the Dalton Highway and Trans-Alaska Pipeline from the northern end of Galbraith Lake to Slope Mountain. The reverse side shows detailed vegetation maps of the 20-square-kilometer (7.7-square-mile) research area centered on Toolik Lake and a 1.2-square-kilometer (1/2-square-mile) intensive research grid on the south side of Toolik Lake. All the maps are part of a hierarchical geographic information system and the Web-based Arctic Geobotanical Atlas.

Courtesy of Alaska Geobotany Center, Institute of Arctic Biology, University of Alaska, Fairbanks.

Vegetation in the Vicinity of the Toolik Field Station, Alaska

This map represents the bathymetry of the Cabonga reservoir, located in the réserve faunique La Vérendrye at the border of Abitibi-Témiscamingue and Outaouais. It was designed to help fishermen and other boating enthusiasts to navigate the waters safely.

The TRAK survey team collected the data over several weeks. Then data was sorted, processed, and checked by the geomatics department with the help of local partners. In order to make the map more versatile, several service providers around the reservoir are also listed.

Courtesy of TRAK and Base de données topographiques du Québec.

TRAKMAPS
Saint-Donat, Québec, Canada
By Sophie Gagné

Contact
Faraj Nakhleh
fnakhleh@trakmaps.com

Software
ArcGIS Desktop 9.2

Printer
Epson 10600

Data Sources
TRAK bathymetric survey, Québec Topographic Database topographic data

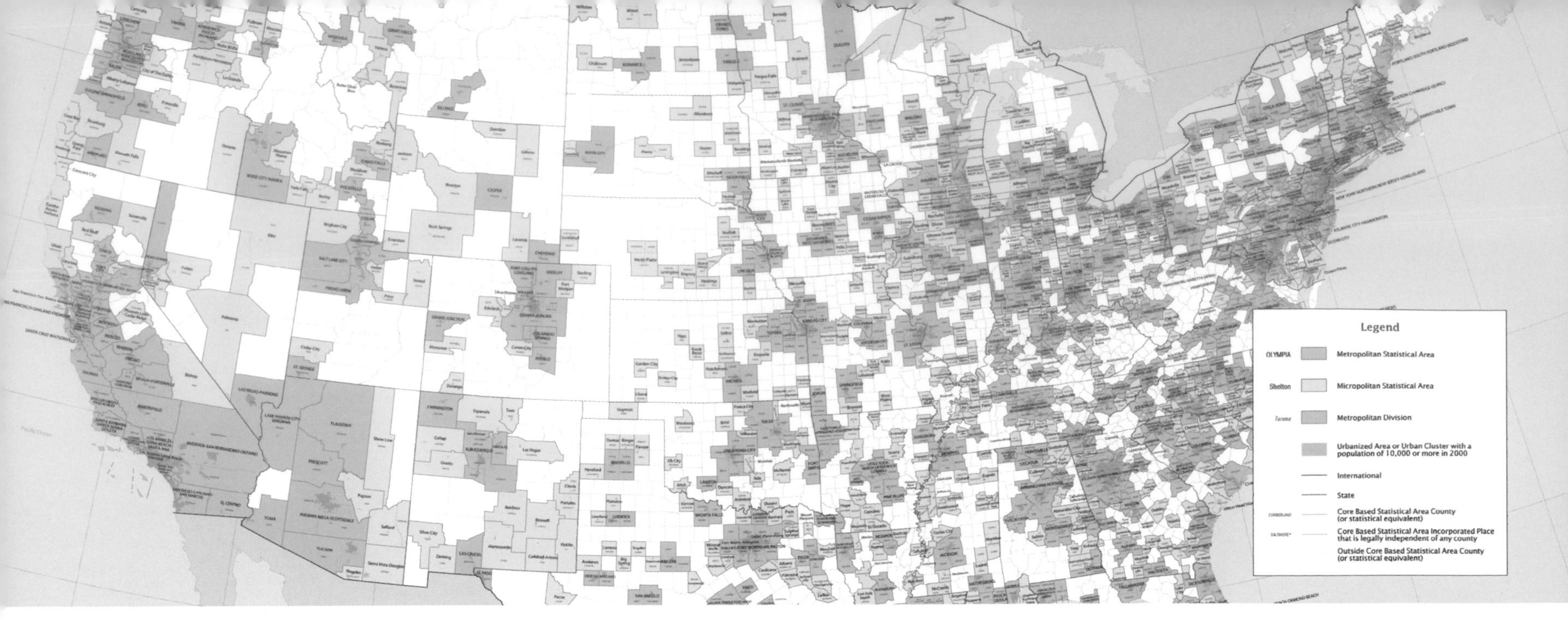

U.S. Census Bureau, Geography Division
Washington, D.C., USA
By Stephanie Spahlinger

Contact

Connie Beard
constance.beard@census.gov

Software

ArcInfo Workstation

Printer

HP Designjet 5000

Data Sources

Office of Management and Budget Statistical Area definitions issued in November 2007, and the U.S. Census Bureau TIGER geospatial data as of 2002

The Metropolitan and Micropolitan Statistical Areas of the United States and Puerto Rico wall map (top) shows metropolitan and micropolitan statistical areas (also known as core-based statistical areas), shows metropolitan divisions, and identifies their component counties. The Combined Statistical Areas of the United States and Puerto Rico wall map (bottom) shows combined statistical areas, and identifies their component metropolitan and micropolitan statistical areas. The printed map measures 55.5-by-36 inches. The boundaries and titles depicted on this map reflect the definitions issued by the Office of Management and Budget (OMB) in November 2007. The area definitions are based on the application of the 2000 Standards for Defining Metropolitan and Micropolitan Statistical Areas to Census Bureau population estimates for incorporated places and selected minor civil divisions for 2005 and 2006, and in specified circumstances, local opinion.

Courtesy of the U.S. Census Bureau, Geography Division.

Statistical Areas of the United States and Puerto Rico, November 2007

Each decade, as part of its tabulation and publication activities following the decennial census, the U.S. Census Bureau calculates the country's center of population. The center is determined as the place where an imaginary, flat, weightless, and rigid map of the United States would balance perfectly if all residents were of identical weight. For Census 2000, the mean center of population was in Phelps County, Missouri, approximately 2.8 miles east of the rural community of Edgar Springs.

Historically, the movement of the center of population has reflected the expansion of the country, the settling of the frontier, waves of immigration, and migration west and south. Since 1790, the center of population has moved steadily westward, angling to the southwest in recent decades.

Courtesy of the U.S. Census Bureau, Population Division.

U.S. Census Bureau
Washington, D.C., USA
By U.S. Census Bureau, Population Division

Contact
Marc Perry
marc.j.perry@census.gov

Software
ArcGIS Desktop 8.3 and 9.1, Adobe Illustrator

Data Sources
Mean centers of population 1790 to 2000 from U.S. Census Bureau, Geography Division

Center of Population, 1790–2000

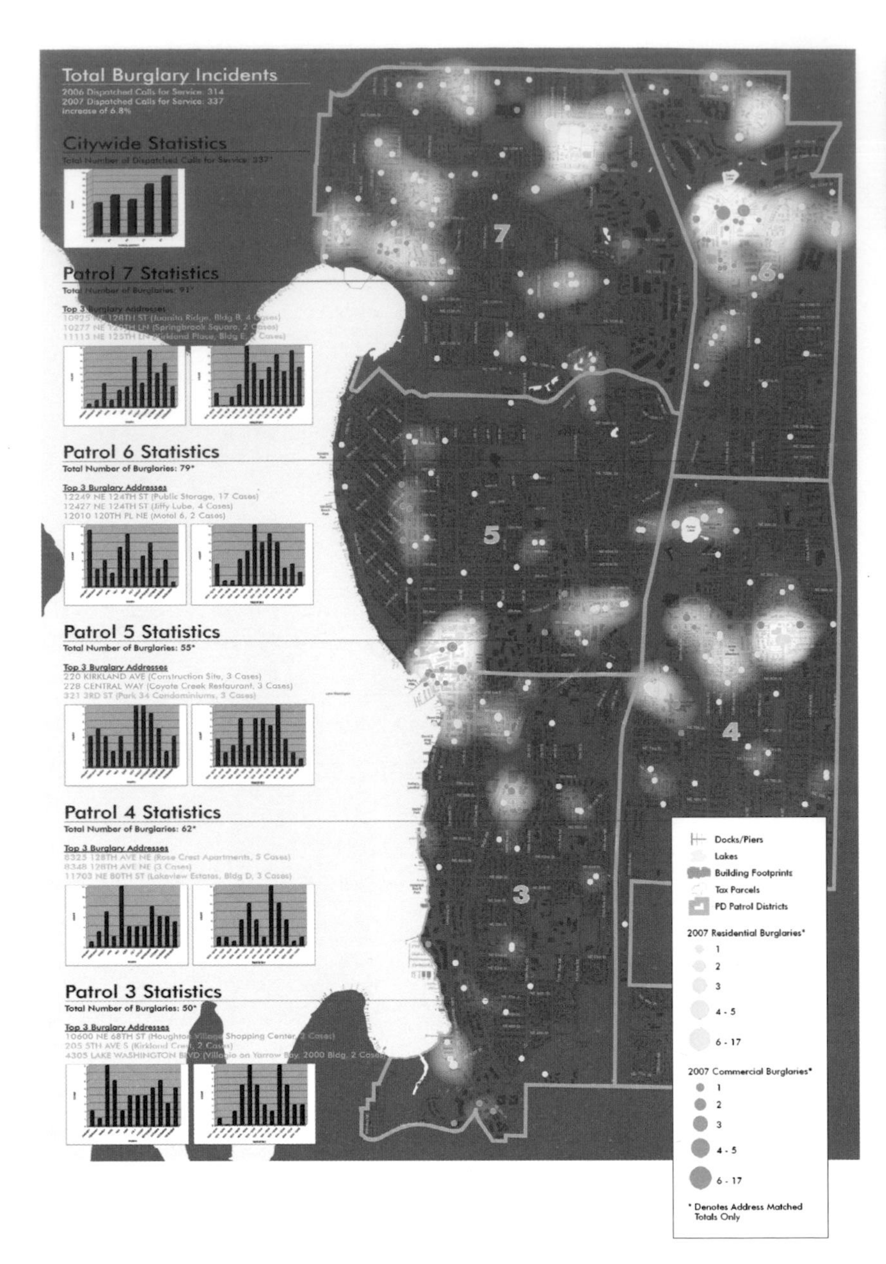

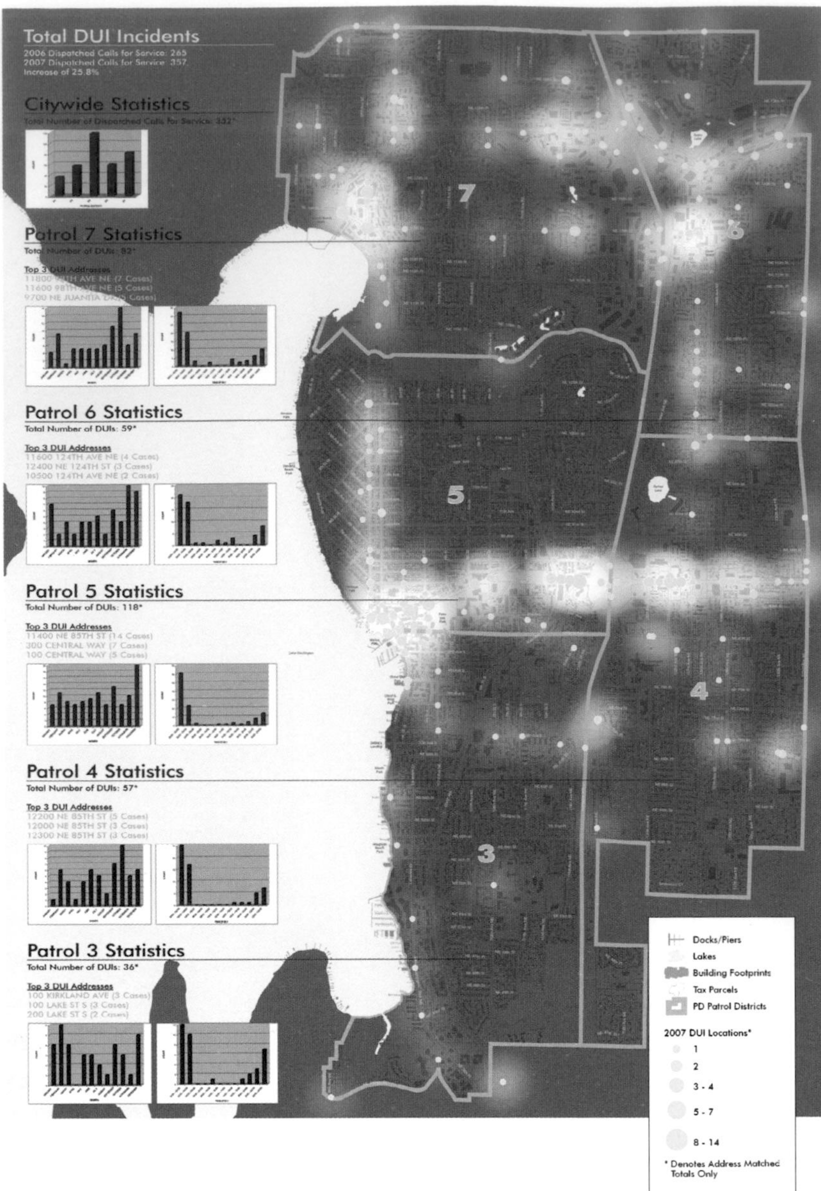

City of Kirkland
Kirkland, Washington, USA
By Christopher Mast

Contact
Christopher Mast
cmast@ci.kirkland.wa.us

Software
ArcGIS Desktop 9.2, ArcGIS Spatial Analyst

Printer
HP Designjet 5000ps

Data Sources
City of Kirkland, King County GIS

Every January, the City of Kirkland produces a series of maps that show the crime trends for the previous year. Six maps are created in total, including burglaries, driving under the influence (DUI) arrests, vehicle prowls, vehicle thefts, traffic collisions, and dispatched calls for service.

The raw data is exported from the dispatch system into a Microsoft Excel spreadsheet. At this point, the data is formatted and converted to a database file. The addresses are geocoded against the city's address point layer or street centerlines (depending on incident type). Any unmatched addresses are then matched manually.

Using ArcGIS Spatial Analyst, the (kernel) density of the matched addresses is calculated. The density raster is then added to the map to illustrate, with just a quick glance, areas where incidents are happening.

Charts summarizing the data for each patrol district (beat) are then created using Excel so the officers can see what is happening in their district graphically. These charts are imported into ArcGIS and placed in the final map layout.

2008 Crime Map Series

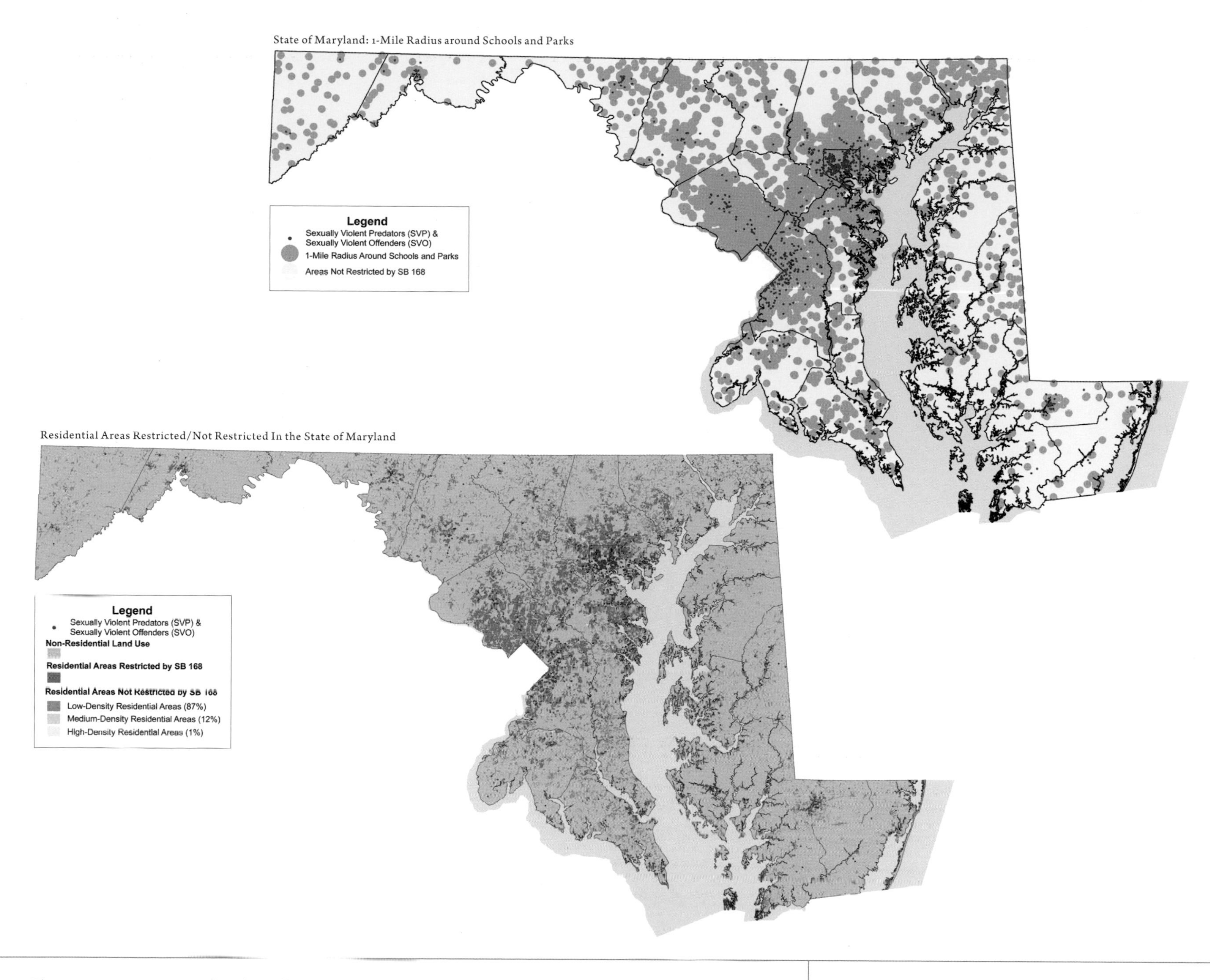

These maps represent a spatial analysis of the impacts of residency restrictions on sexual offenders resulting from 2007 state legislation in Maryland. The analysis included areas in Maryland where registered sex offenders would and would not be able to live, the number of sex offenders who would be required to move, and the current residential housing densities (low, medium, high) within and outside of areas that would be restricted to sex offenders.

The radius map shows a 1-mile radius around schools and parks in Maryland, along with the location of sexually violent predators and sexually violent offenders on Maryland's Sex Offender Registry. The 1-mile radii are created from a center point of each location, not from the perimeter of each school campus or park. The off-limits areas for offender residencies will be larger than shown. The parks do not include any state parks, but do include many community and residential parks. The schools include public schools and universities and colleges, but do not include many private schools. It is important to note that off-limit areas for offender residencies will be larger than shown with additional data of school and park locations.

The residential restrictions map shows generalized residential areas in Maryland that are either restricted or not restricted for sex offenders due to legislation. The analysis concluded that 33.4 percent of residential land is not restricted to sex offenders for residency. Of that land, 87 percent is low-density residential, 12 percent is medium-density residential, and 1 percent is high-density residential.

Courtesy of Erin Lesh and Ashley Lesh Buzzeo.

Erin Lesh and Ashley Lesh Buzzeo
Millersville, Maryland, USA
By Erin Lesh and Ashley Lesh Buzzeo

Contact
Erin Lesh
elesh17@gmail.com

Software
ArcGIS Desktop 9.2 and Adobe Photoshop

Printer
HP Designjet 5500 ps

Data Sources
U.S. Census, ESRI, MDP, Registry Web site

Sex Offenders and Housing: Analyzing Implications of Housing Restrictions on Registered Sex Offenders

Collin County

Mckinney, Texas, USA

By Tim Nolan, Nick Enwright, Bret Fenster, Kendall Holland, Gabriela Voicu, Ramona Luster, Larry Krieger, and Adil Abdalla

Contact

Gabriela Voicu

gvoicu@co.collin.tx.us

Software

ArcGIS Desktop 9.2, ArcGIS Network Analyst 9.2

Printer

HP Designjet 5500

Data Sources

City of Wylie, Collin County

Since 2005, Collin County has had a local agreement with the City of Wylie to provide GIS services for the Wylie Fire and Rescue Department. Specific fire district GIS layers were created from historical computer-aided design (CAD) drawings and integrated into the Collin County database. After the base layers were complete, spatial analysis became possible.

This map shows the average of actual emergency call response times between June 2006 and August 2007. The data was prepared by interpolating geocoded response times. The interpolated surface was then smoothed using neighborhood statistics. Accuracy of the surface layer increases in areas with a greater number of calls.

Courtesy of Collin County GIS.

Wylie Fire and Rescue Actual Response Times

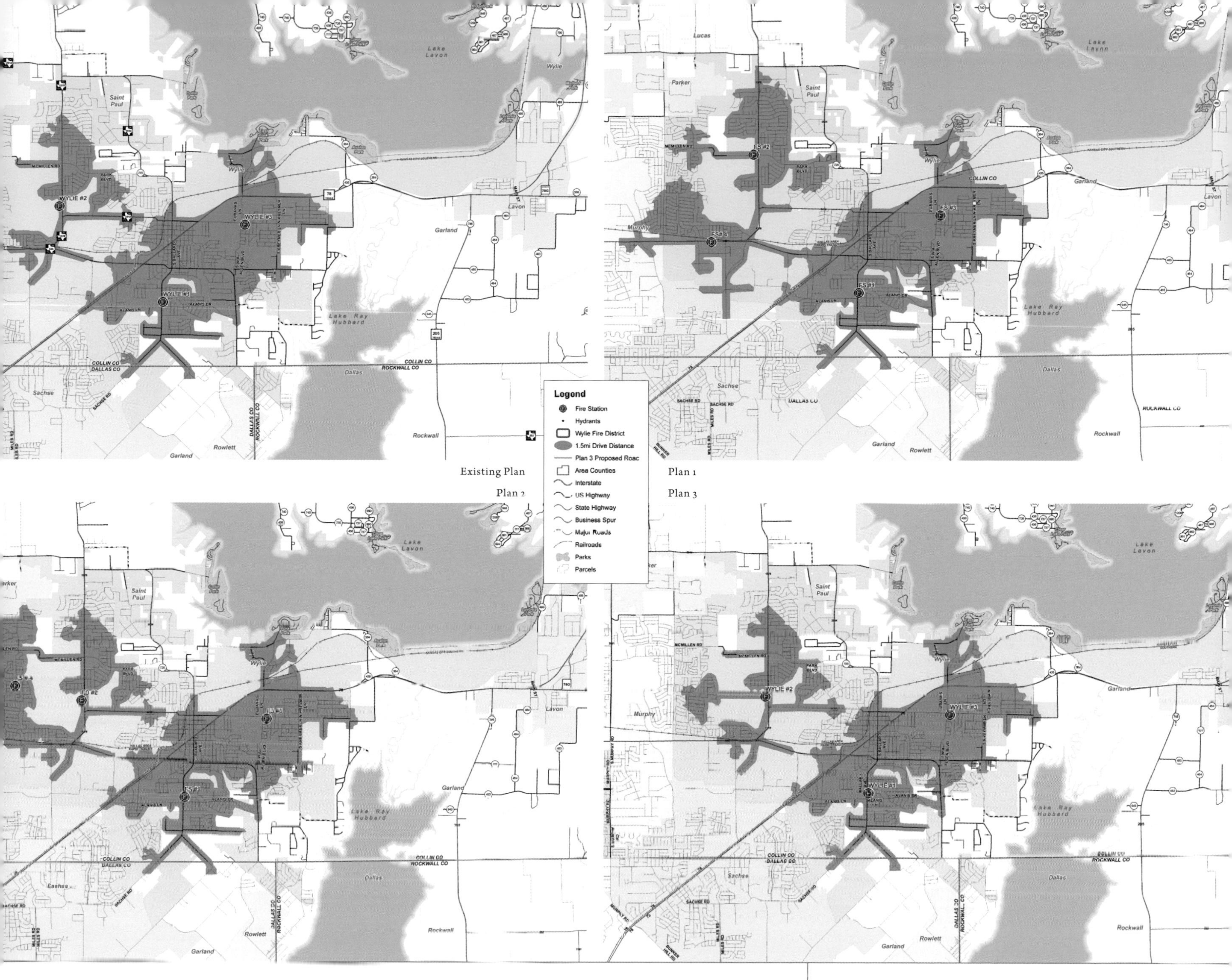

The Insurance Services Office (ISO) rating of an area is closely associated to the insurance premiums assessed within that specific area. In this case, the rating reflects the resources of the City of Wylie and its Fire and Rescue Department.

The Wylie Fire and Rescue Department requested that Collin County GIS validate its current Standard Response District (SRD). The task involved network analysis to calculate travel distances for fire engines. The study sought to examine Wylie's needs and plan for the placement of additional fire stations, which would extend services, accommodate future growth, and maintain Wylie's highest rating of ISO (1).

ArcGIS Network Analyst was used to delineate the SRD, which is composed of 1.5-mile drive distance buffers for each fire station. The map includes four plans: the current plan and three suggested plans. The proposed plans suggest either adding or moving a station, and adding a new road. Each plan changes the SRD area and thus the number of hydrants in the SRD. The goal of this analysis was to assist and support Wylie in its decision while following the recommendations and requirements outlined in Hickey's Fire Suppression Rating Schedule Handbook. The best choice is the plan that serves the highest number of people (parcels) and includes the greatest number of hydrants in the SRD.

Courtesy of Collin County GIS.

Collin County
Mckinney, Texas, USA
By Tim Nolan, Nick Enwright, Bret Fenster, Kendall Holland, Gabriela Voicu, Ramona Luster, Larry Krieger, and Adil Abdalla

Contact
Gabriela Voicu
gvoicu@co.collin.tx.us

Software
ArcGIS Desktop 9.2, ArcGIS Network Analyst 9.2

Printer
HP Designjet 5500

Data Sources
City of Wylie, Collin County

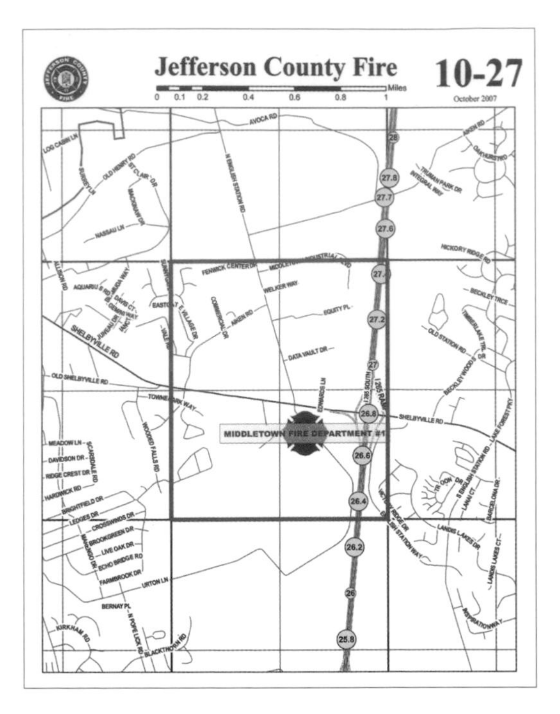

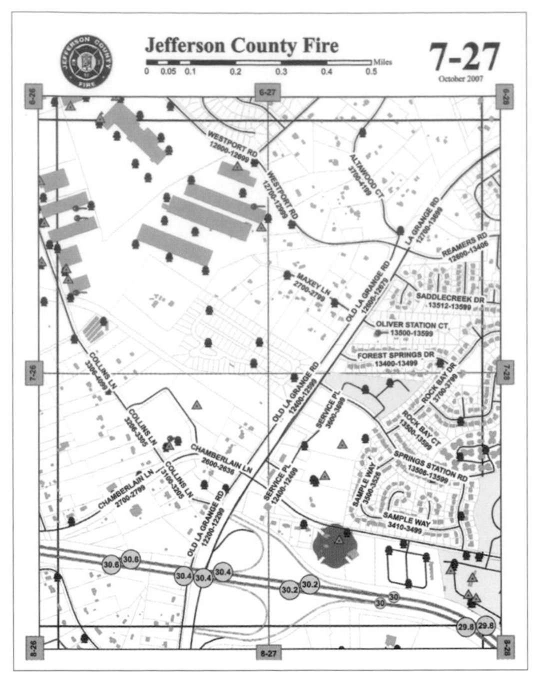

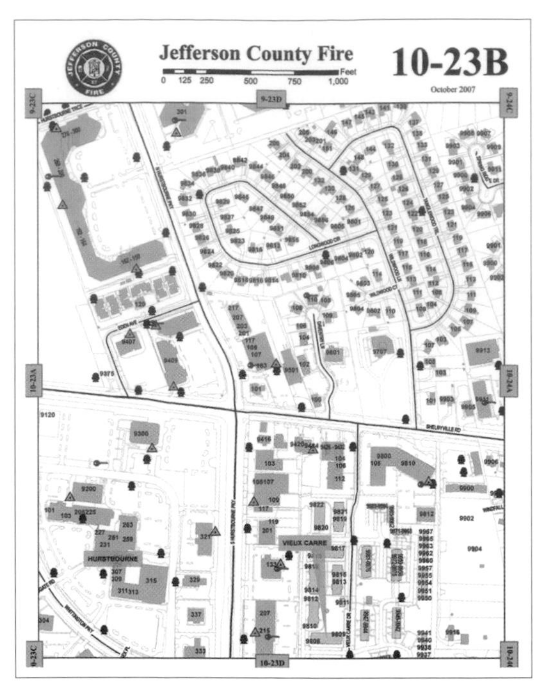

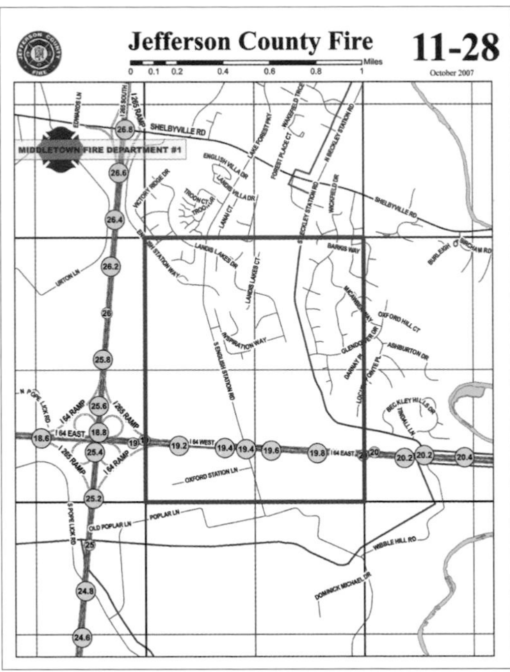

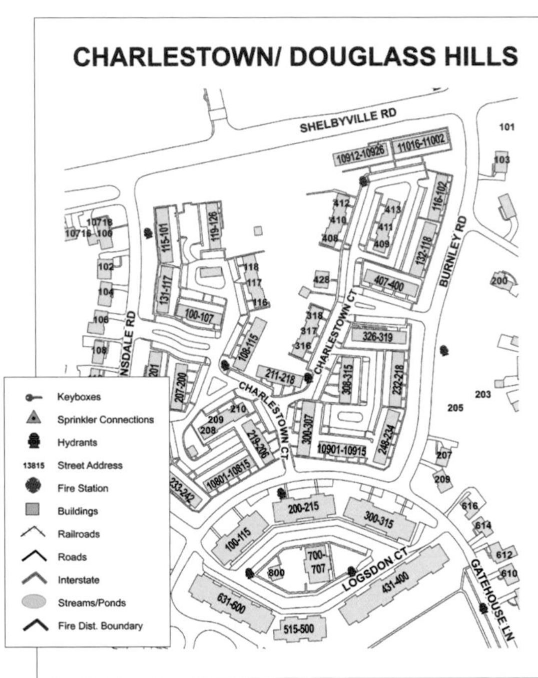

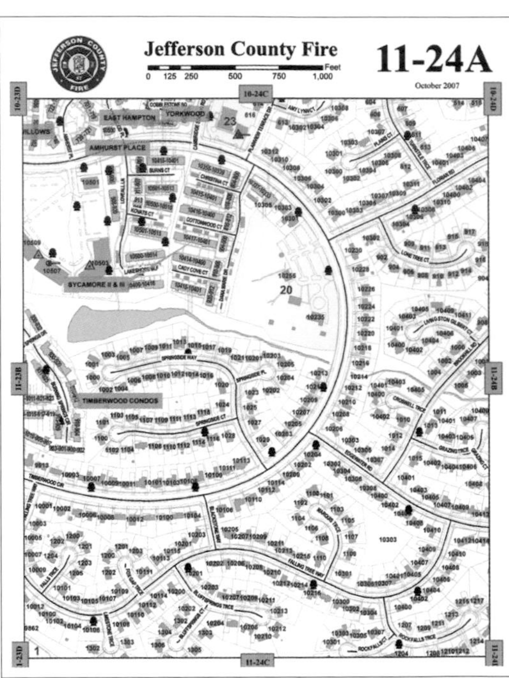

Middletown, Kentucky, Fire Atlas

Louisville and Jefferson County Information Consortium for Middletown Fire Protection District
Louisville, Kentucky, USA
By Andy Longstreet and Angela Scott

Contact
Jane Poole
poole@lojic.org

Software
ArcGIS Desktop 9.2, Mapbook

Printer
Ricoh Aficio CL7100

Data Source
Louisville and Jefferson County Information Consortium for Middletown Fire Protection District

This atlas is used by the Middletown Fire Protection District as a first responders guide. In January 2003, the City of Louisville and Jefferson County merged into one government body, the Louisville/Jefferson County Metro Government. Residents inside the boundaries of the former City of Louisville and the suburban service area are protected by separate fire districts. The suburban area of Jefferson County spans more than 300 square miles and is serviced by over 1,100 volunteer and career firefighters.

Fire personnel use map books in their vehicles to help them efficiently respond to calls. Each district is responsible for updating the books that cover its jurisdiction. The old map book areas were unique, based on physical features that cover large variances of area, and contained outdated information, especially in the rapidly developing eastern section of Jefferson County.

In 2006, the Middletown Fire Protection District automated atlas production and updating, designing a new grid system that would eventually cover all of Jefferson County, cross fire district boundaries, have an alphanumeric reference, accommodate 8 ½-by-11-inch page printing, and be based on Louisville and Jefferson County Information Consortium data.

After Middletown adopted the new grid system in 2006, six neighboring fire districts started using it. These fire districts are also working together to map private hydrants, keyboxes, and sprinkler connections, which are shown in the atlas.

Courtesy of Louisville and Jefferson County Information Consortium.

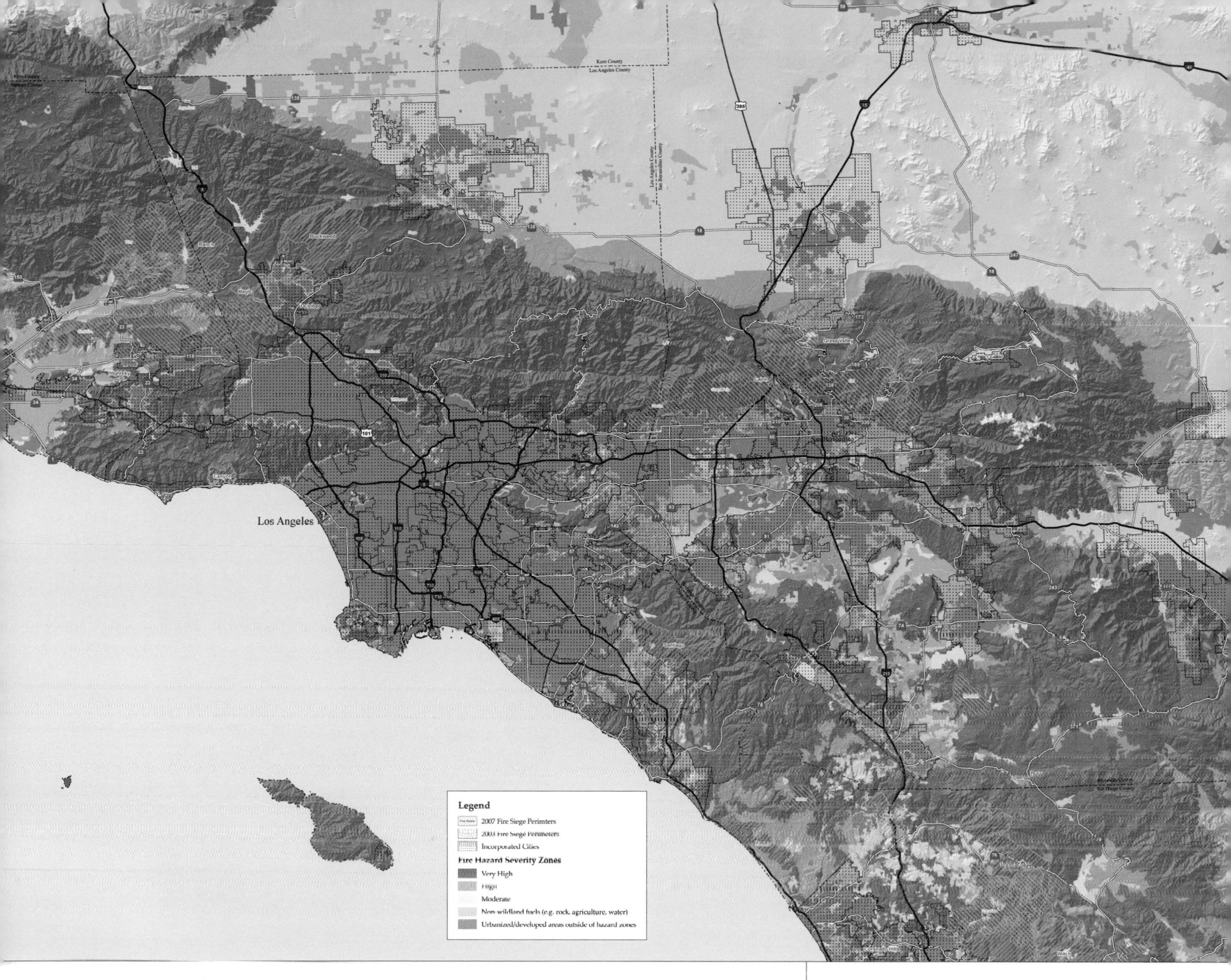

State law directs the California Department of Forestry and Fire Protection (CAL FIRE) to map areas of significant wildfire hazards. CAL FIRE's Fire and Resource Assessment Program created the map of draft Fire Hazard Severity Zones (FHSZ) using data and models describing development patterns, potential fuels over a 30- to 50-year time horizon, expected fire behavior, and expected burn probabilities to quantify the likelihood and nature of vegetation fire exposure (including firebrands) to new construction.

This map provides a generalized validation of the FHSZ model results by comparing the model's zones to fire perimeters from the 2007 and 2003 fire sieges. The draft FHSZs are also being improved through a more exhaustive and detailed validation process involving CAL FIRE field staff, local government, fire protection districts, and the public.

Courtesy of California Department of Forestry and Fire Protection, Fire and Resource Assessment Program.

California Department of Forestry and Fire Protection, Fire and Resource Assessment Program
Sacramento, California, USA
By Ron Caluza

Contact
Ron Caluza
ron.caluza@fire.ca.gov

Software
ArcGIS Desktop 9.2

Printer
HP Designjet 1055 cm Plus

Data Sources
CAL FIRE Fire and Resource Assessment Program, California Spatial Information Library, local government agencies

Draft Fire Hazard Severity Zones with 2007 and 2003 Fire Siege Perimeters: Southern California

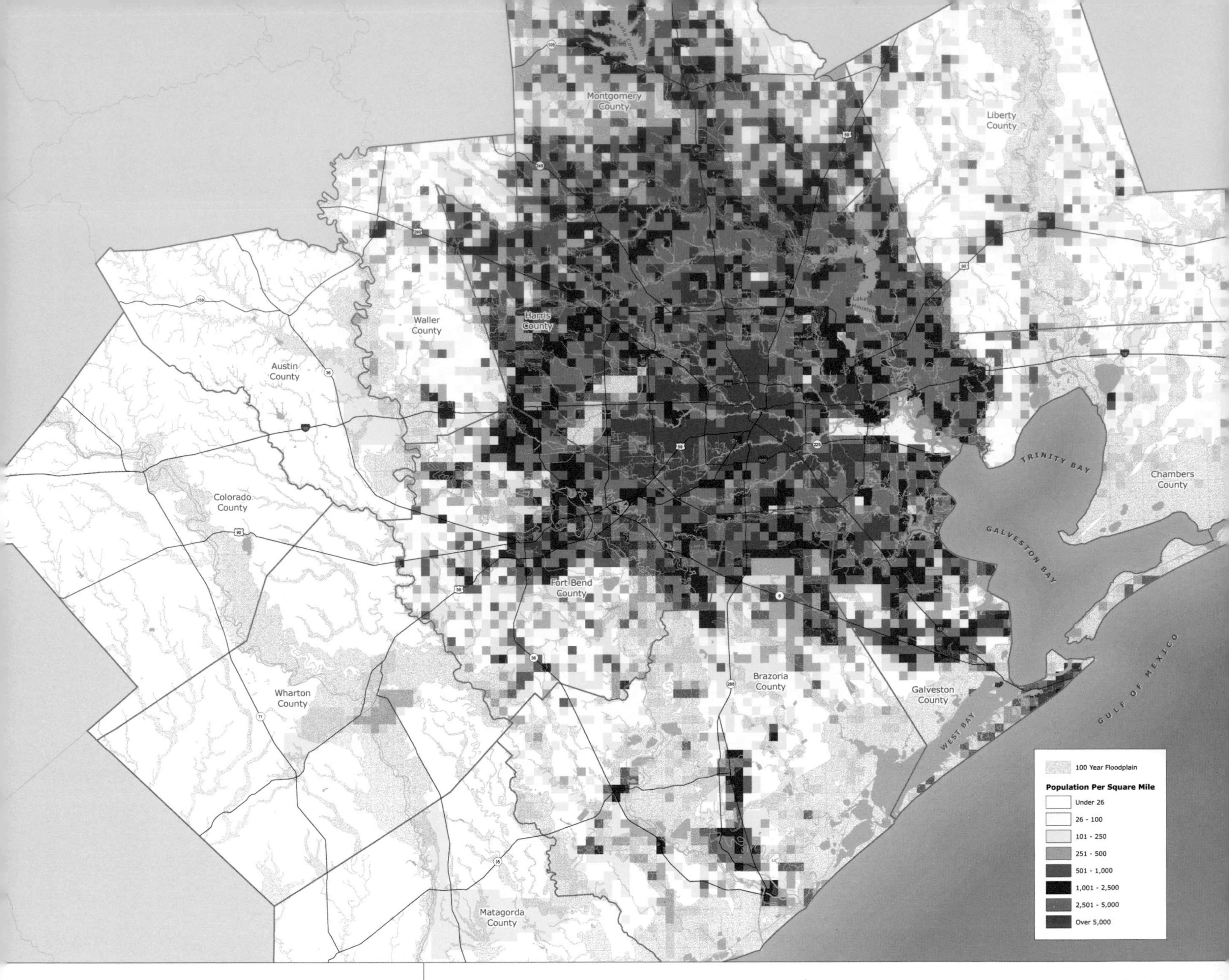

Houston-Galveston Area Council
Houston, Texas, USA
By William Bass, GISP

Contact
William Bass
william.bass@h-gac.com

Software
ArcGIS Desktop 9.2, SAS 9.1

Printer
HP Designjet 1055cm

Data Sources
Houston-Galveston Area Council, Texas Natural Resources Information System, U.S. Census, Federal Emergency Management Agency

Over five million people live in the Houston-Galveston region today, and the population is projected to grow to over 8.8 million by 2035. A region of this size is continually evolving and developing to support the growing population.

The Houston-Galveston region also contains over 4,500 square miles of floodplains. The intersection of people and floodplains produces a mixture of environmental and socioeconomic risks for the region. The analysis for this project was performed as part of a regional global climate change and flood hazards initiative.

The analytical methods involved a combination of ArcGIS spatial analysis tools as well as analytical and data manipulation methods in SAS. The integration of ArcGIS and SAS allowed for a large amount of data to be gathered and processed. This map was one in a series that represented the culmination of the project's efforts.

13-County Houston-Galveston Region: Population Inside the 100 Year Floodplain (2035)

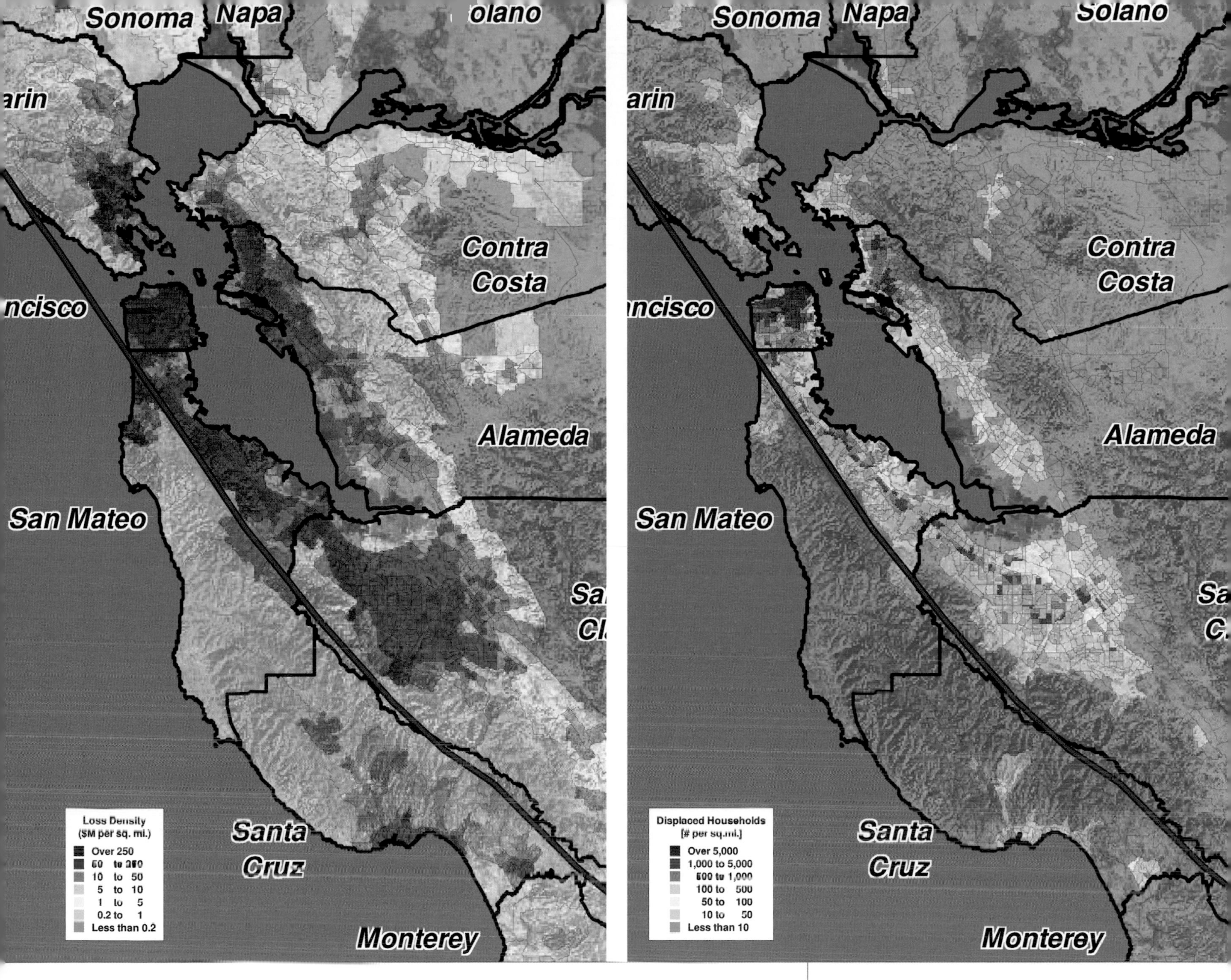

Economic Loss

Hazards U.S. Multi-Hazard (HAZUS-MH), the risk-assessment software program, was used in a study of building damage and losses likely to occur in a repeat of the 1906 San Francisco earthquake. Depicted in this map are direct economic losses from a possible recurrence of that catastrophic event.

These losses include capital stock losses (repair and replacement costs of the structural system, the nonstructural system, and building contents), and income losses (business interruption, temporary rental space and moving costs, and other expenses relating to building function due to structural system damage).

Displaced Households

This map shows estimates of displaced households from a repeat of the 1906 event, which is a function of the number of residences in the nineteen-county study region that would experience either extensive or complete structural damage. HAZUS-MH also estimates short-term shelter requirements, taking into account the income levels of the impacted population. These HAZUS-MH products are very useful in assessing potential short-term shelter and long-term housing requirements following a major earthquake.

Courtesy of FEMA Mitigation Directorate.

Federal Emergency Management Agency Mitigation Directorate
Washington, D.C., USA
By Federal Emergency Management Agency

Contact
Melis Mull
melis.mull@dhs.gov

Software
ArcGIS Desktop, HAZUS-MH, Adobe PhotoShop CS3

Printer
HP Laserjet 9000

Data Sources
HAZUS-MH, State of California

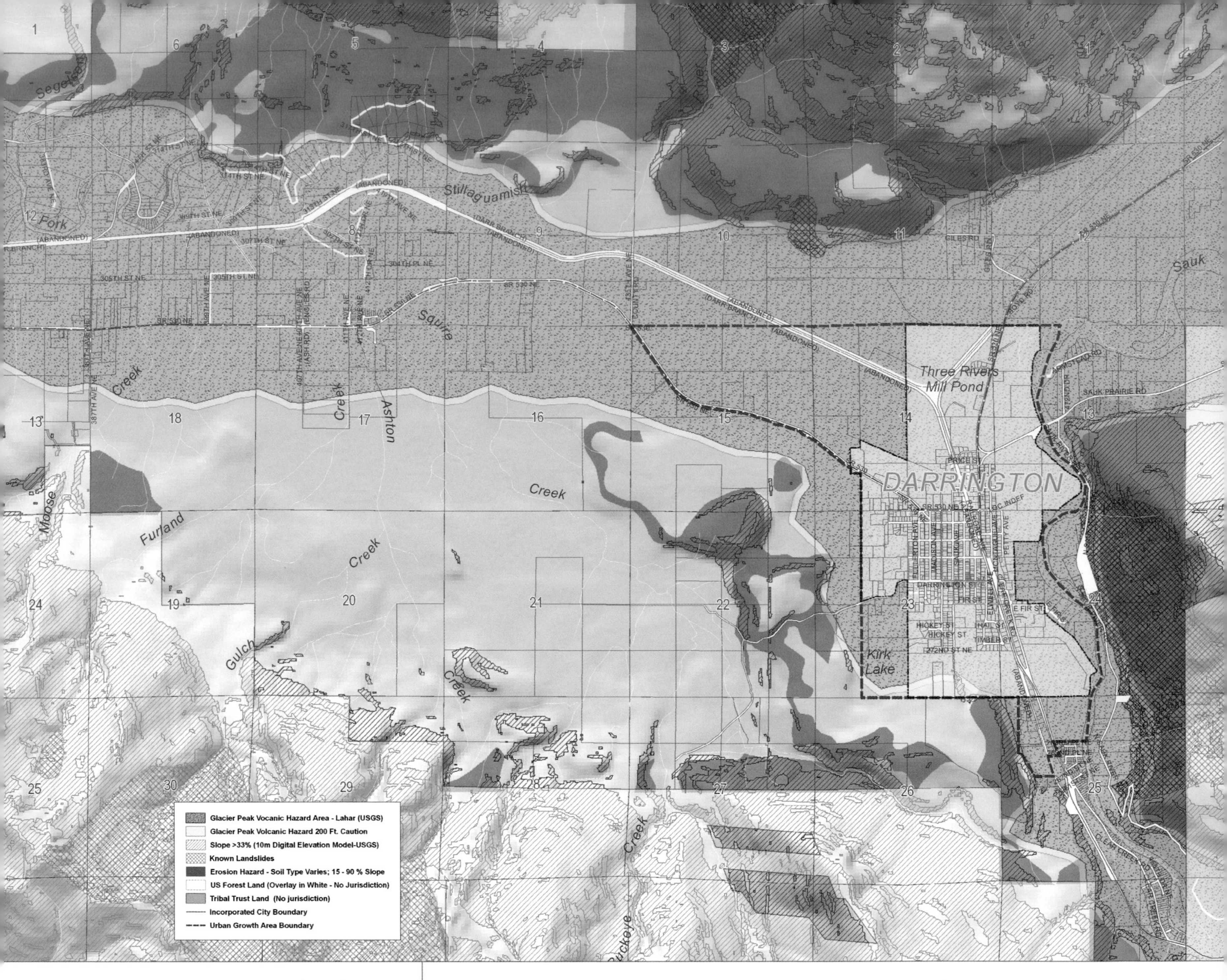

Snohomish County
Everett, Washington, USA
By Department of Planning and Development Services; Flynn Adams

Contact
Carrol Lane
cb.lane@co.snohomish.wa.us

Software
ArcGIS Desktop

Printer
HP Designjet 1055cm

Data Sources
Washington State Department of Natural Resources, U.S. Geological Survey, Snohomish County

Snohomish County, in compliance with the Washington State Growth Management Act, adopted Critical Area Regulations in 2007. This was to provide protection for critical resources such as wildlife, wetlands, and water supplies.

This volcanic hazards map is part of the Critical Areas Landslide, Erosion, and Volcanic Hazards Atlas. The atlas was created using ArcGIS software with the PLTS extension, incorporating data from file-based coverages, shapefiles, and ArcSDE geodatabases. Data layers for the fourteen-township map atlas include lahar (debris flow) hazard areas, erodible surficial geology, landslides, and slopes greater than 33 percent.

Courtesy of Snohomish County Department of Planning and Development Services.

Snohomish County Critical Areas Landslide, Erosion, and Volcanic Hazards—Township Series Map Sheet 7

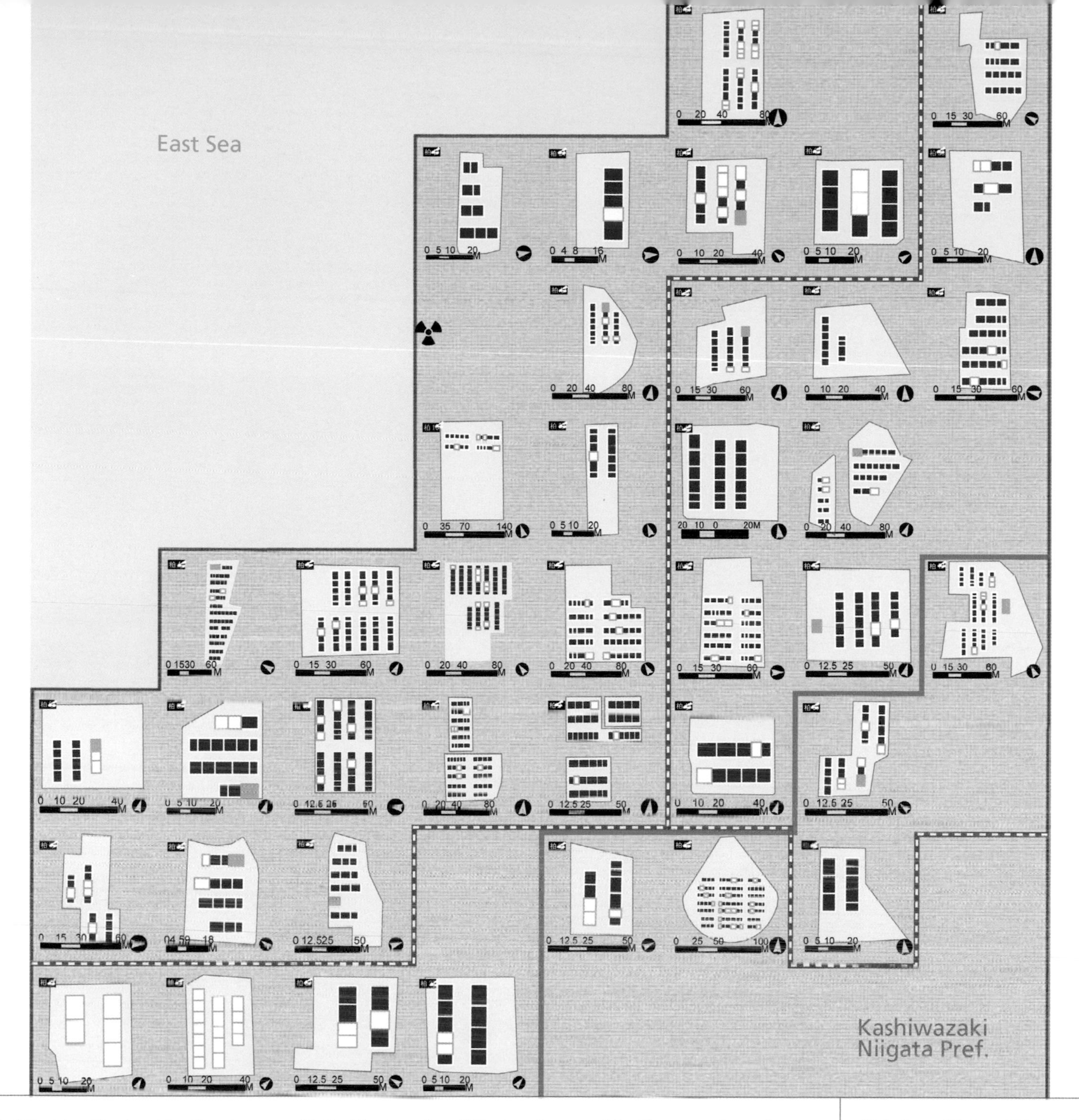

Occupied
Move-out / Vacant
Dayroom
Japan Railway
HOKURIKU Expwy

The Niigataken Chuetsu-oki earthquake occurred in the Chuetsu region of Niigata prefecture on July 16th, 2007, and caused a great deal of structural damage. This map describes not only the relative position of the temporary housing site in a disaster-stricken area—grasping the big picture—but also shows specific information of each unit, such as its condition, stored in an attribute table.

The quake refugees are allowed to live in the provisional houses free of charge for up to two years. That means local officials must work effectively to help put people's lives back in order. The lesson learned from past seismic disasters is to pay attention to the people who could end up dying by being isolated in temporary housing.

The housing opportunity started in August 2007. This map represents conditions as of 2008, after the turn of the fiscal year. The occupancy ratio decreased by 84.3 percent at the time. In these situations, georeferential relational database management systems, such as building damage reports and the basic resident register, were corrected by local officials and emergency response research teams. They were used to support local officials in the rehabilitation and reconstruction period.

Courtesy of Yokohama National University, Kyoto University, Niigata University, and Niigata GIS Association.

Yokohama National University, Center for Risk Management and Safety Sciences
Yokohama, Kanagawa, Japan
By Takashi Furuya, Haruo Hayashi, Go Urakawa, Kanehisa Fujiharu, Keiko Tamura, Munenari Inoguchi, and Hiroko Sakai

Contact
Takashi Furuya
t-furuya@ynu.ac.jp

Software
ArcGIS Desktop 9.2, Adobe Illustrator CS2

Printer
Epson PX-9000

Data Sources
The City of Kashiwazaki in Niigata prefecture, Chuo Group

Deformed Overhead View Map for Distribution and Occupancy of Temporary Housing Unit with Time-Series Variation after the Niigataken Chuetsu-oki Earthquake in 2007

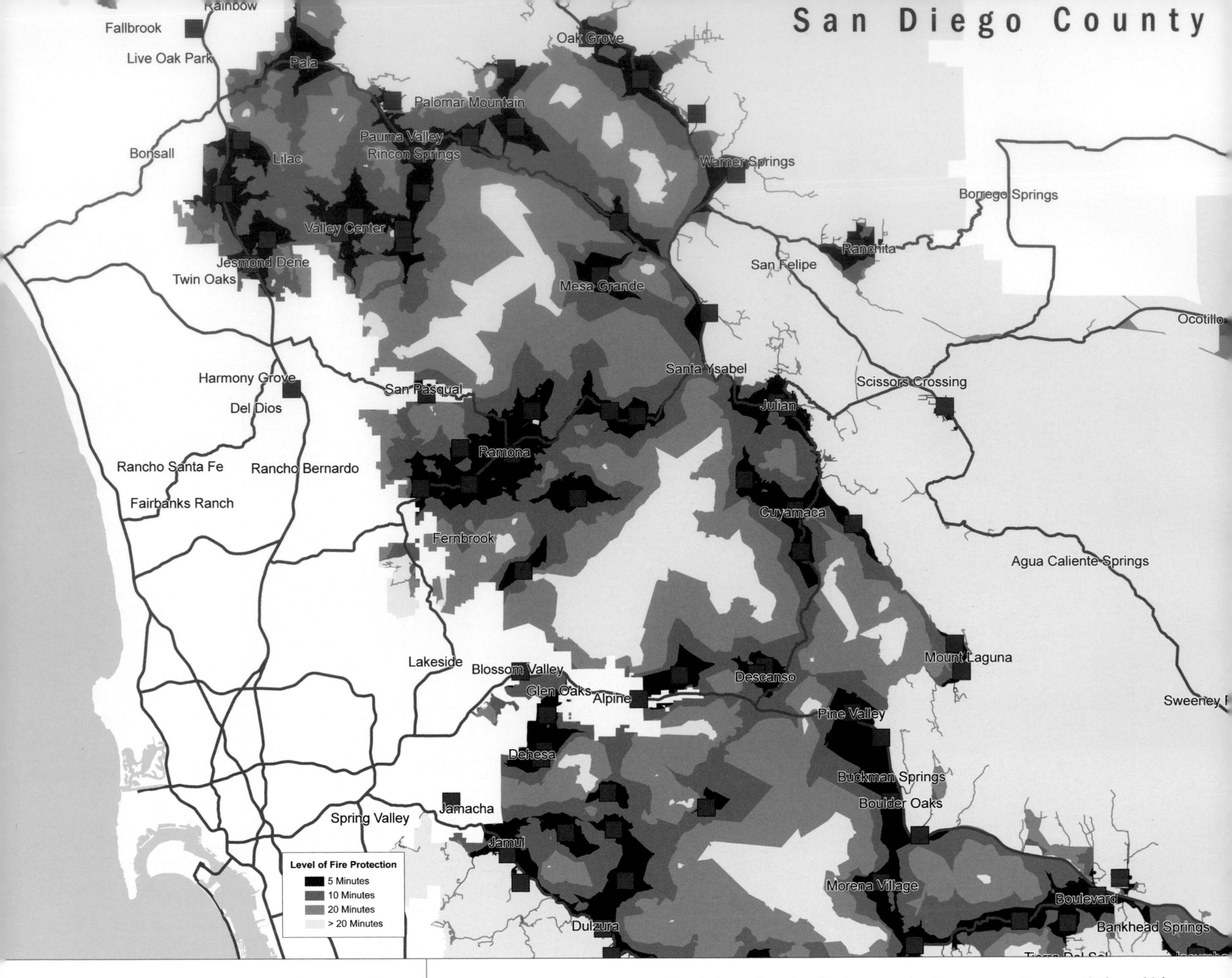

County of San Diego/GIS
San Diego, California, USA
By Orson Bevins

Contact
Orson Bevins
orson.bevins@gmail.com

Software
ArcGIS Desktop 9.2, ArcGIS Network Analyst

Printer
HP Designjet 1055cm Plus

Data Sources
SanGIS, City of San Diego Fire-Rescue

The County of San Diego has historically relied on independent fire districts and volunteer companies to provide fire and life safety protection. As a result of major wildfires in recent years, county officials have pursued a reorganization plan to create an administrative body that will efficiently coordinate response to emergencies, thereby improving service. This map shows approximate response times from existing fire stations at full implementation of the plan. It was used in a report to the Local Agency Formation Commission, which oversees administrative boundary changes.

An effective fire protection strategy will incorporate the premise that most people and assets are clustered near transportation infrastructure. That idea raised a question: What are the number and proportion of dwellings and other structures of value that can be reached in specific time frames? To help understand the answer, an analysis was prepared using a roads feature class from the County of San Diego Department of Public Works, the county assessor parcel dataset, and a node dataset from the City of San Diego. Some jurisdictions were not analyzed as shown by the off-white color. The ArcGIS Network Analyst extension was used to calculate the geometry of each response time polygon. The number and proportion of dwellings and structures were obtained by selecting those parcels with an improvement value greater than zero, whose geographic center lay within a specific response time polygon. Areas in gray were excluded. The results of the analysis demonstrate that, with proper coordination, emergency response from existing fire stations can improve service to these areas of the county.

San Diego County Parcel-Based Fire Protection Analysis

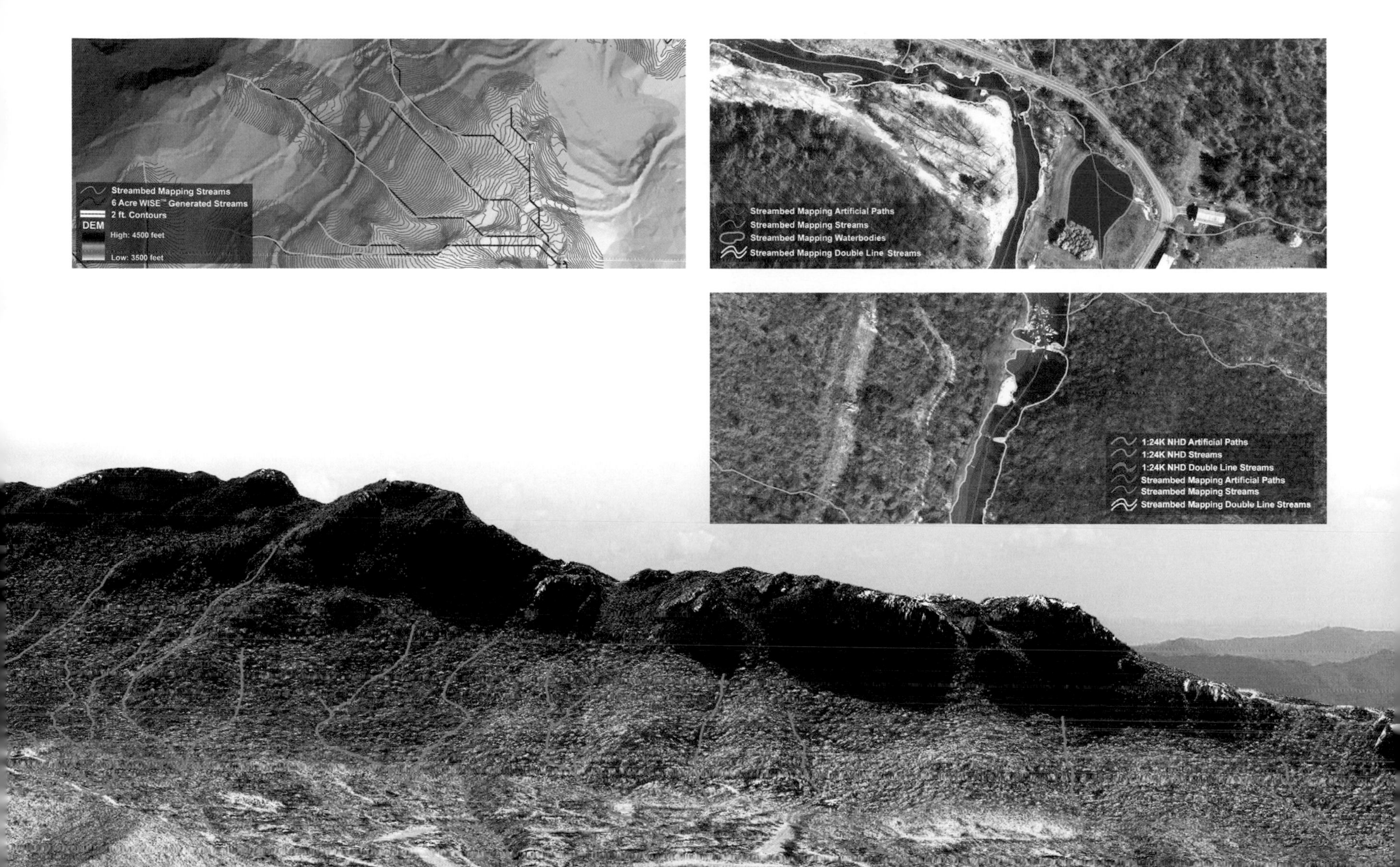

In the fall of 2004, Hurricanes Ivan and Frances battered significant portions of North Carolina, leaving nineteen counties in the western region declared federal disaster areas. In response, the North Carolina General Assembly, through Senate Bill 7, developed the Hurricane Recovery Act. The act created several programs to support the response and recovery of these areas. This package included support for the North Carolina Stream Mapping Project to execute the recommendations of the implementation plan. Under the direction of The North Carolina Center for Geographic Information and Analysis (CGIA), this high-resolution stream mapping dataset was completed in August of 2007.

AECOM Water worked closely with CGIA and an advisory committee composed of personnel from multiple federal, state, and local municipalities to accurately assess all user needs for the stream mapping dataset. This committee helped to establish user requirements, a database schema design, a six-acre drainage area requirement for the termination of the streamlines, and a quarter-acre size requirement for the collection of water bodies.

Courtesy of AECOM Water.

AECOM Water
Greensboro, North Carolina, USA
By John Hendricks

Contact
John Hendricks
john.hendricks@aecom.com

Software
ArcGIS, ArcScene, Adobe Photoshop

Printer
HP Designjet 1055cm Plus

Data Sources
2.5m post spaced lidar, U.S. Geological Survey 1:24,000 National Hydrography Dataset (NHD) data, 0.5m ground pixel resolution imagery

Western North Carolina Streambed Mapping Project

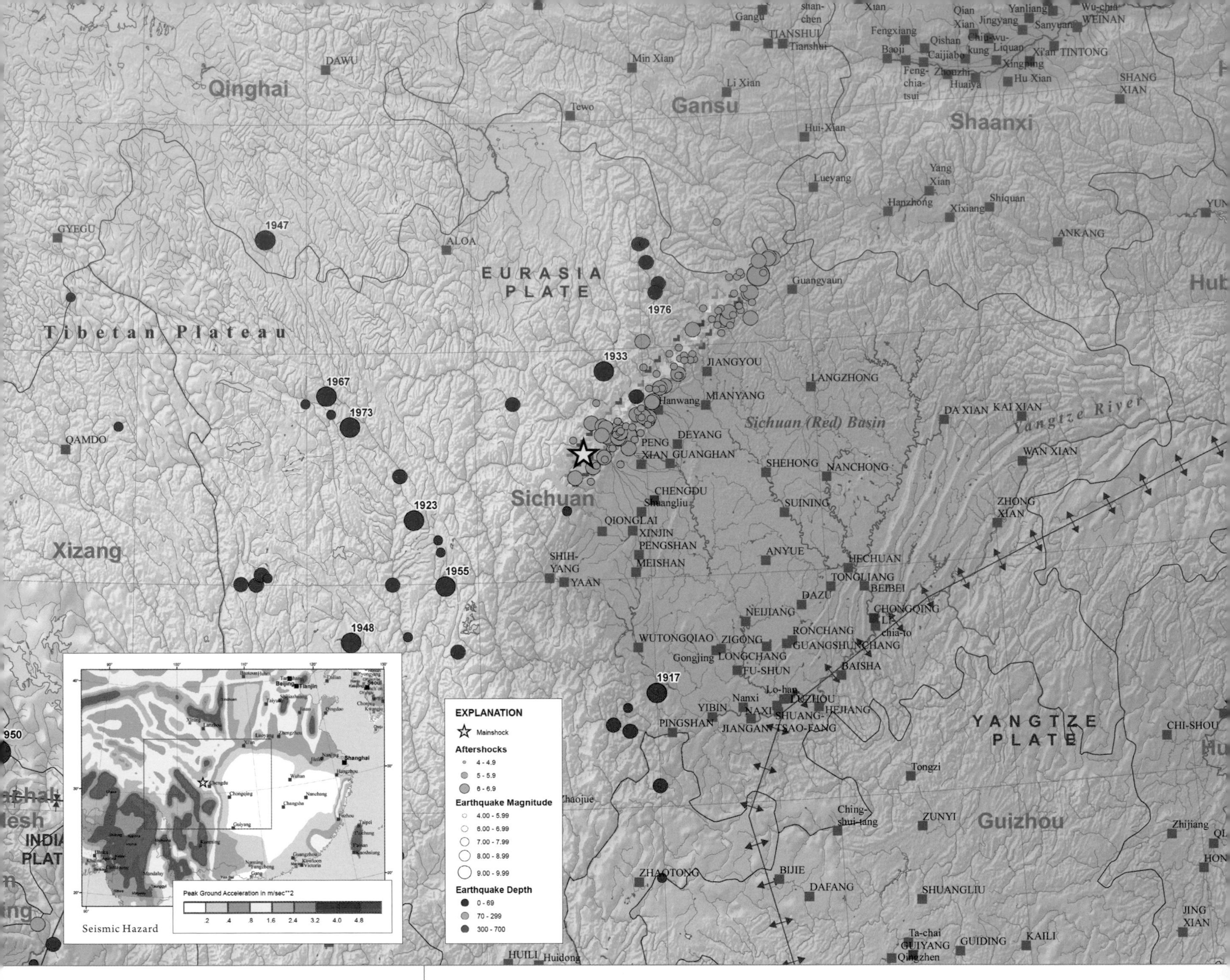

U.S. Geological Survey
Golden, Colorado, USA
By Susan Rhea

Contact
Susan Rhea
rhea@usgs.gov

Software
ArcGIS Desktop

Printer
HP Designjet 5550

Data Source
U.S. Geological Survey

The Earthquake Hazards Program of the U.S. Geological Survey uses ArcGIS software to produce earthquake summary posters of significant earthquakes and special study maps of seismic hazards in the United States and elsewhere. These maps are used for a range of needs, from congressional presentations and college geology classes to city planning.

The poster of the M7.9 May 2008 China event concisely shows the impact of this devastating earthquake. Seismotectonic history compares the event with other large earthquakes. A close up of the epicentral region puts the event in cultural context, and the seismic hazard map puts the event in long-range planning context. Additional figures inform emergency responders and seismologists studying large earthquake rupture, and provide a historical reference.

The Indonesia seismic hazard poster is used in preparing the local government agencies and the U.S. foreign aid agencies for future risk in the country. The earthquake summary map shows areas affected by great earthquakes before the 2004 tsunami earthquake and seismicity since then. Areas of concern are those that are not yet showing post-tsunami earthquakes, as history has shown that large earthquakes will occur there sometime. This poster helps tell the story of why earthquake preparation needs to be an ongoing effort throughout Indonesia.

Courtesy of the U.S. Geological Survey.

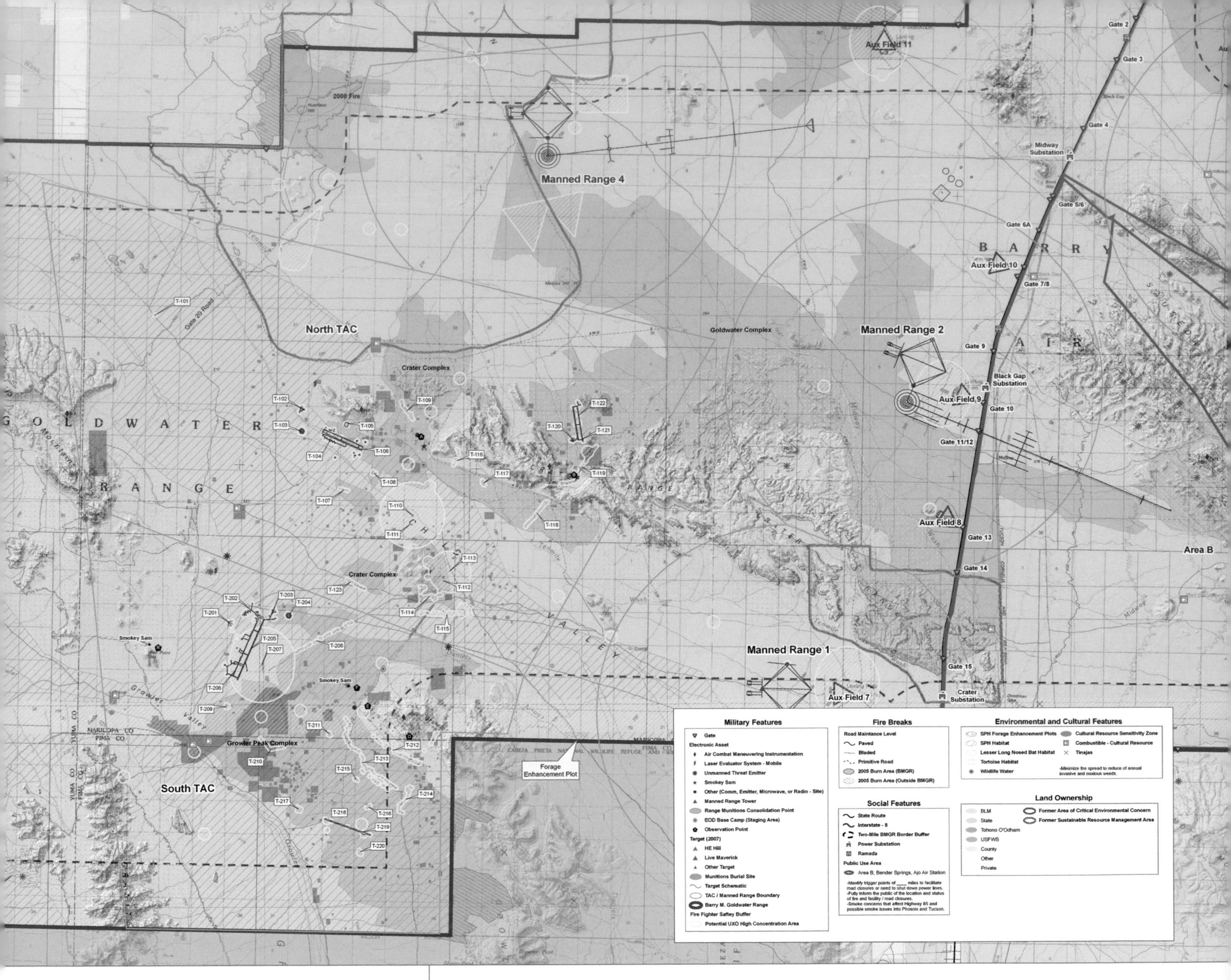

U.S. Air Force, Luke Air Force Base, 56 Range Management Office
Luke Air Force Base, Arizona, USA
By Chris Black

Contact
Chris Black
chris.black@luke.af.mil

Software
ArcGIS Desktop 9.2

Printer
HP Designjet 1055

Data Sources
U.S. Geological Survey, U.S. Bureau of Land Management, U.S. Air Force, Arizona Game and Fish Department

This map was developed to support the wildland fire plan for the 1.05-million-acre Barry M. Goldwater Range—East. It provides planners and decision makers a view of the spatial relationships of important features in the consideration of wildland fire, including military assets and sensitive environmental and cultural resources.

The map shows factors relevant to fire ignition and spread—such as previously burned areas, fire breaks, and topography. It provides wildland fire responders with critical information, including firefighter safety zones, areas where unexploded ordnance is likely to exist, location and condition of roads, position of high-value electronic equipment, and location of combustible archeological sites. The map also includes the fire management response flowchart and contact numbers for neighboring land managers and fire responders.

Courtesy of Luke Air Force Base.

Barry M. Goldwater Range—East: Wildland Fire Plan

October 2007 brought nine active wildfires to the San Diego County Region at one time. The California Department of Forestry and Fire Protection (CAL FIRE) maintains a database of fire history for San Diego County dating back to 1910. Every summer, this database is updated with the previous year's fire perimeters.

Included on these maps are fire perimeters for the October 2007 fires. Advanced scripting techniques made it quick and easy to add the 100 years of map data together and produce a final result.

Both maps incorporate all wildfires and prescribed burns as reported to CAL FIRE from 1910 to 2007.

Courtesy of David Toney, GISP.

U.S. Marine Corps/GEO *Fidelis* West

Camp Pendleton, California, USA

By David Toney, GISP

Contact

David Toney

david.toney@usmc.mil

Software

ArcGIS Desktop 9.2

Printer

HP Designjet 5500 ps

Data Sources

ESRI, San Diego Association of Governments, SanGIS, California Department of Forestry and Fire Protection

San Diego County Fire History

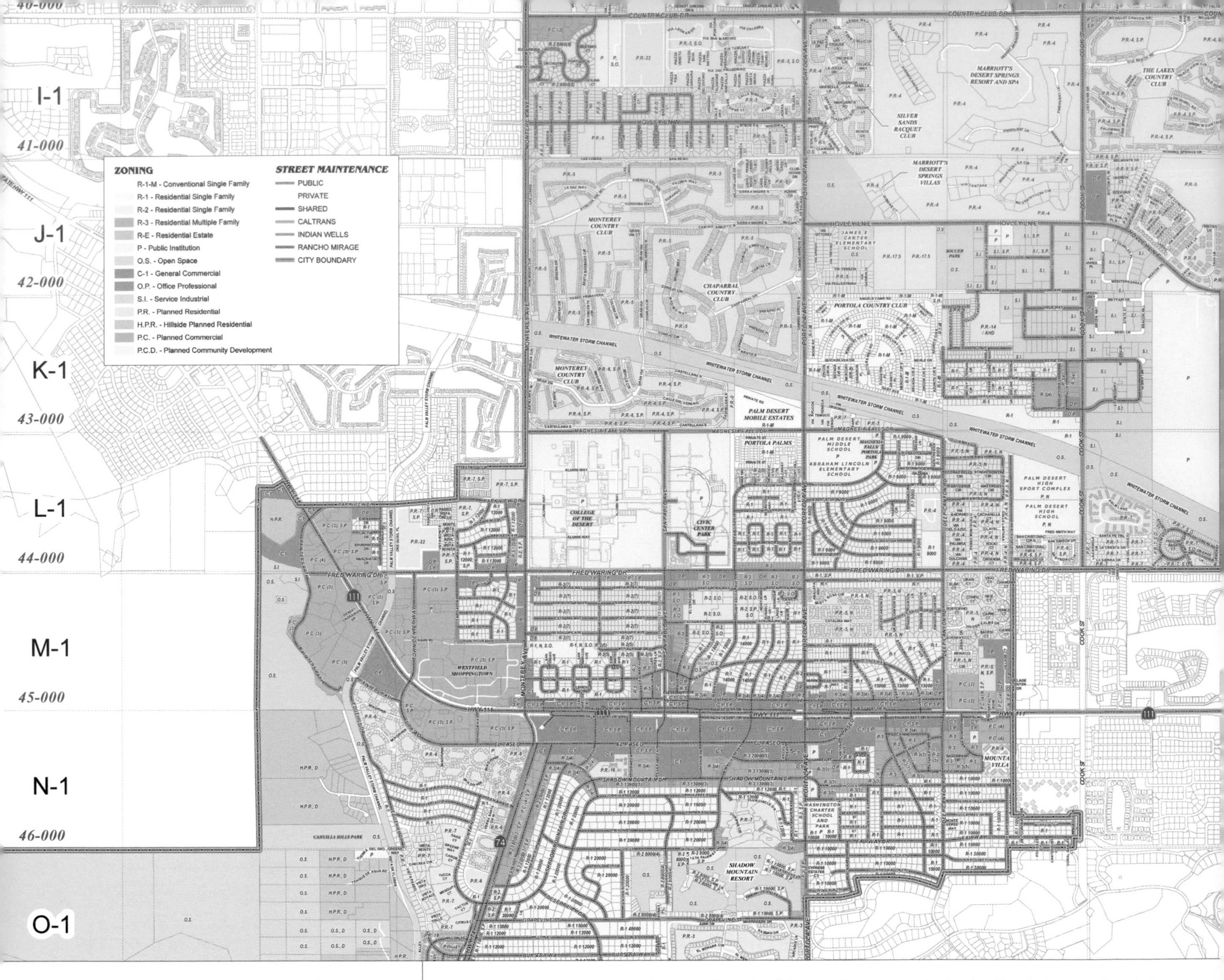

City of Palm Desert

Palm Desert, California, USA
By Bob Riches

Contact
Bob Riches
briches@ci.palm-desert.ca.us

Software
ArcGIS Desktop

Printer
HP Designjet 5000

Data Sources
City of Palm Desert, Riverside County

The City of Palm Desert's zoning map has evolved from a hand-drawn Mylar map, created and updated by a draftsman working in the city's Planning Department, to a digital map. The decision to go digital was made in 1998, and the map has been revised through the years using ArcInfo 7.x, ArcView 3.x, and now ArcGIS 9.3. The city sells the 36-by-48-inch map, along with an accompanying street index printout, to the public, and it has been a high-demand item before Planning Commission and City Council meetings. All the data that is shown on the paper map is available digitally on the city's internal Web site using ArcIMS and soon ArcGIS Server.

The City of Palm Desert is located in the center of the Coachella Valley, south of the Joshua Tree National Monument, and about 120 miles east of Los Angeles. With a population of more than 45,000 in a 27-square-mile area, Palm Desert is a thriving year-round community with the natural beauty and recreational amenities of a resort destination.

Courtesy of the City of Palm Desert, California.

City of Palm Desert Zoning Map

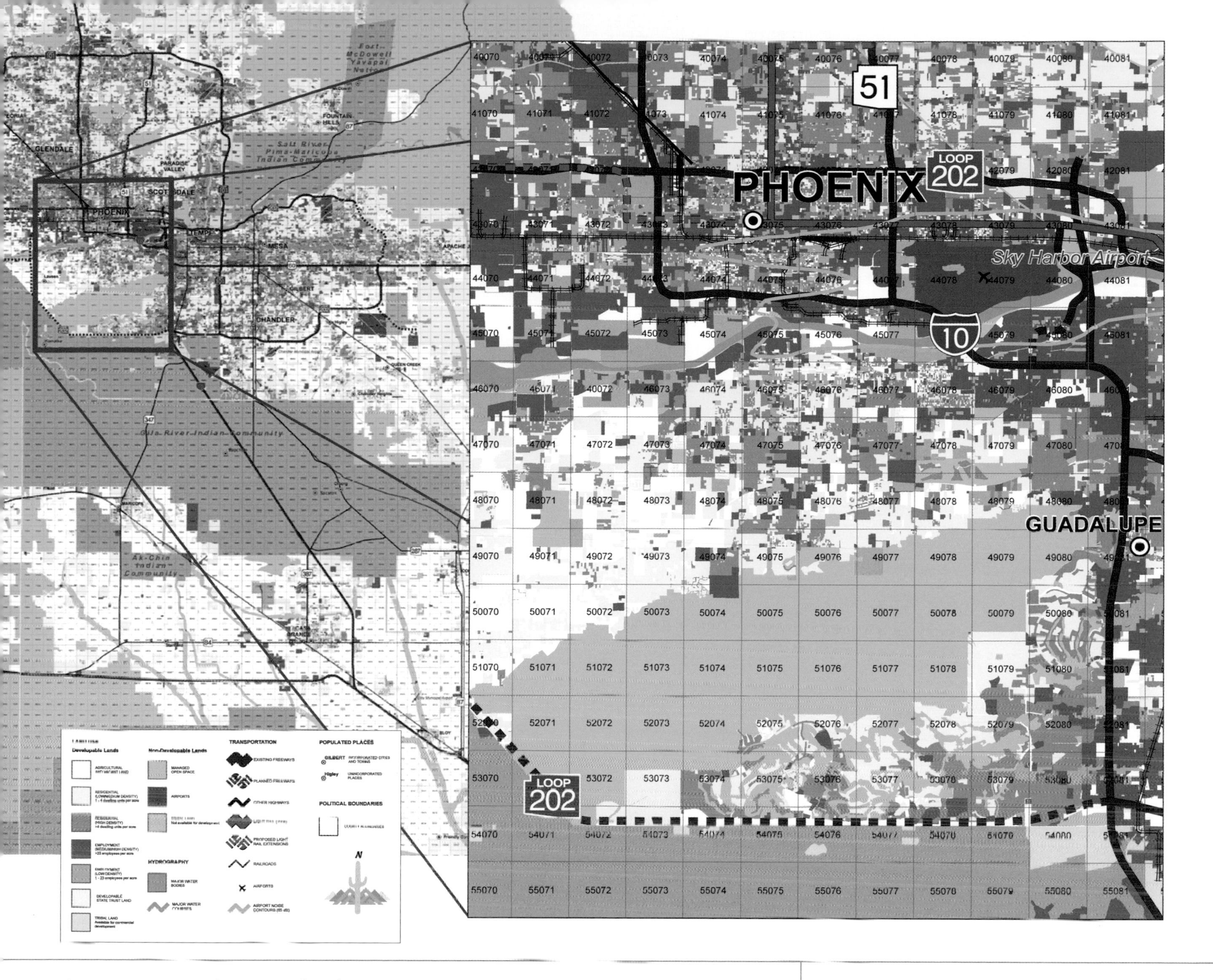

The Maricopa Association of Governments (MAG) partnered with the Urban Land Institute in a regional growth visioning exercise entitled AZ One: A Reality Check for Central Arizona. In the exercise, over 300 participants used Lego blocks to represent the growth of the region by an additional six million people and three million jobs. Legos represented new housing and employment. Maps used by exercise participants in the AZ One event were created by MAG.

The map depicts planned and existing transportation infrastructure, airport noise contours, and existing land use. MAG produced the existing land-use data, which was integrated with similar data produced by the Central Arizona Association of Governments and updated with land divestment information from the U.S. Bureau of Land Management. State Trust land was added as an overlay, indicating areas potentially available for development.

Courtesy of Maricopa Association of Governments.

Maricopa Association of Governments
Phoenix, Arizona, USA
By Anubhav Bagley, Kurt Cotner, Jason Howard, Mark Roberts, and Rita Walton

Contact
Jason Howard
jhoward@mag.maricopa.gov

Software
ArcGIS Desktop 9.2

Printer
HP Designjet 5000 ps

Data Sources
Existing land use: MAG, Central Arizona Association of Governments, and their member agencies; landownership: U.S. Bureau of Land Management, Arizona State Land Department; existing and planned freeways: MAG; existing and planned light rail routes: Valley Metro Rail; airport noise contours: local airports

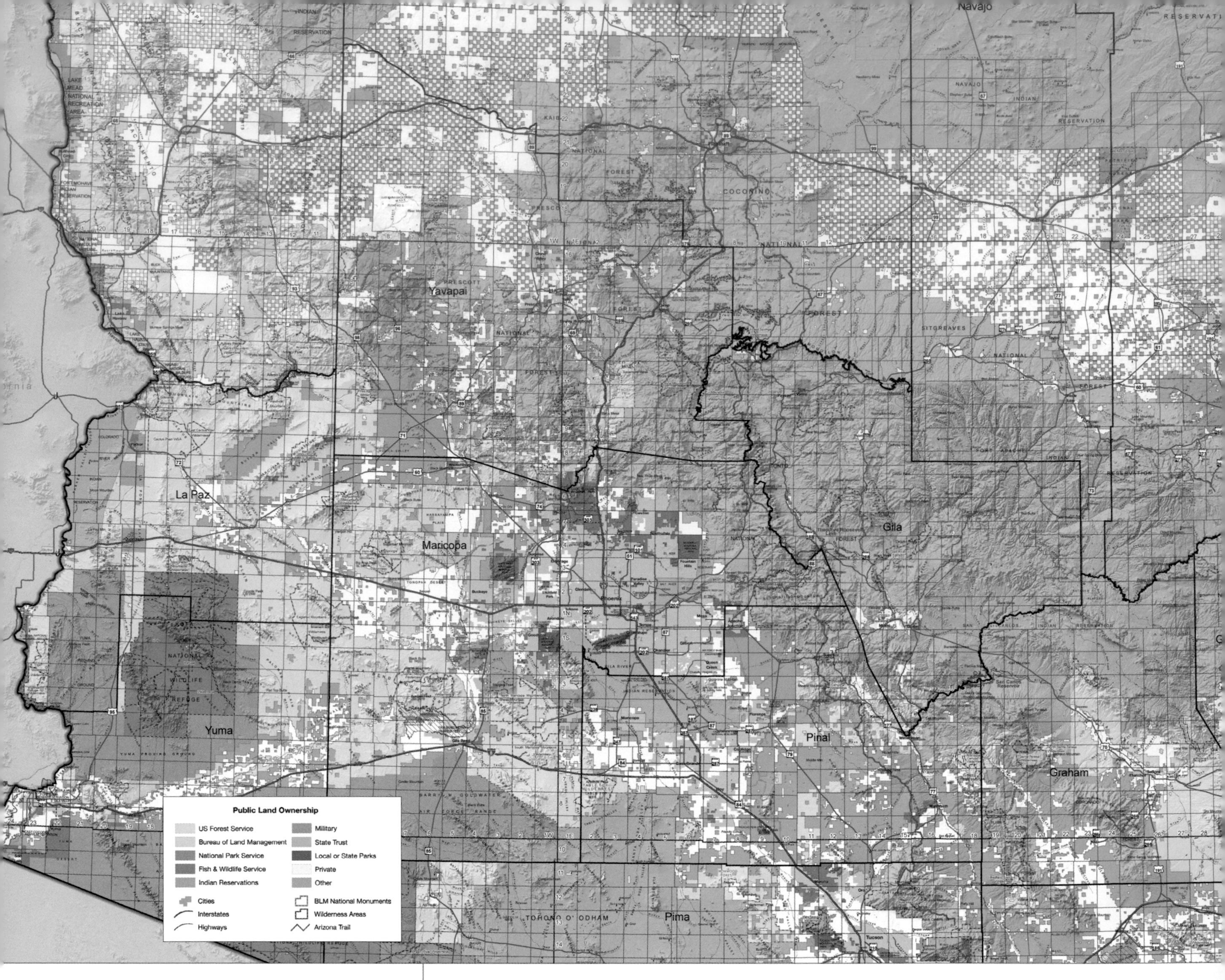

Arizona State Land Department
Phoenix, Arizona, USA
By Ryan Johnson

Contact
Ryan Johnson
rjohnson@land.az.gov

Software
ArcGIS Desktop

Printer
HP Designjet 1055cm

Data Sources
Arizona State Land Department, Arizona Department of Revenue, Arizona Department of Transportation, U.S. Bureau of Land Management, U.S. Census Bureau, Maricopa County, Pima County

Land in Arizona is owned by a wide variety of organizations, including federal, tribal, private, and the state trust. It is important for citizens, businesses, private landowners, and various agencies to know who owns each section of land. The original landownership database was created as a cooperative project between the Arizona State Land Department (ASLD) and the U.S. Forest Service in the 1980s. To facilitate the use of this data, the ASLD developed the Arizona Surface Management Responsibility map series in 1994 using AML scripts to generate the maps.

In 2005, the map series underwent extensive modernization to make use of the enormous cartographic improvements available in ArcGIS software. Today this map series consists of a statewide map, two metro area maps, twelve county maps, and 134 half-degree tile maps. These maps show state, federal, and private landownership, road networks, hydrological features, topography, cities, towns, and points of interest. They have been purchased by a wide variety of customers ranging from governors to school children.

State map

The state map is the centerpiece of the Arizona Surface Management Responsibility map series. It provides a good overview of landownership across the state of Arizona.

Arizona Surface Management Responsibility

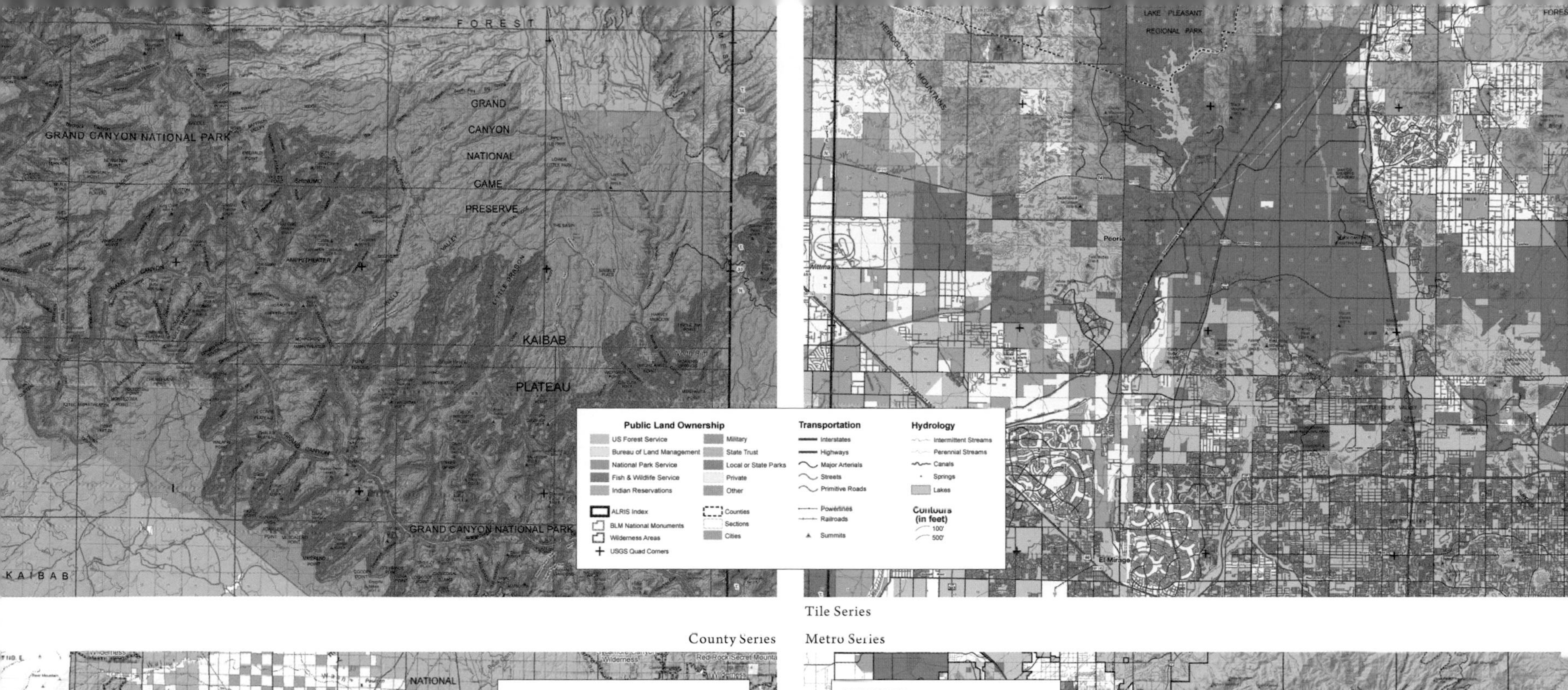

Tile Series

County Series

Metro Series

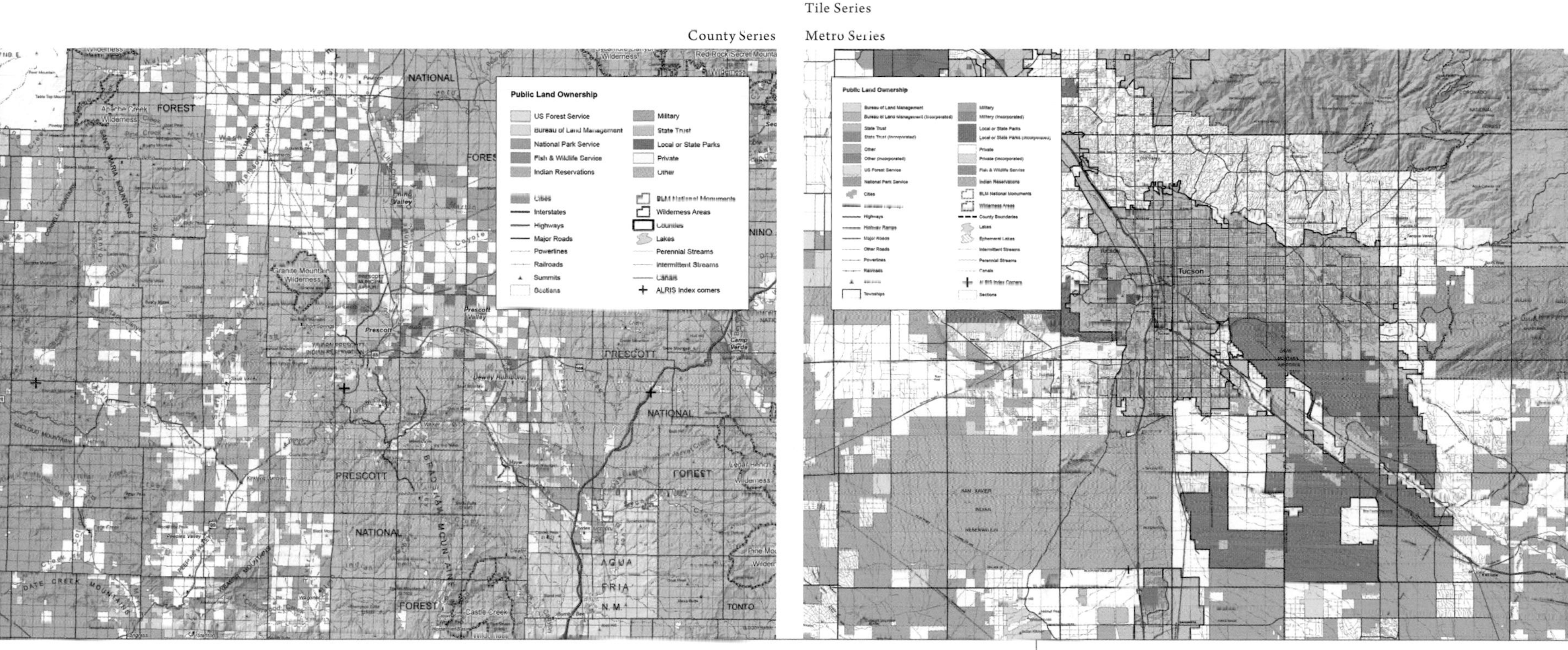

Tile series

Using half-degree arcs, each of these maps is set at a scale of 1:100,000, making them the most detailed of the series. Each map has the same layout with an index that allows users to find a map quickly. These maps are dynamically created by a program developed in-house using ArcObjects. This allows them to be created and exported to PDF in one large batch for quick and frequent updates. These maps cover tiles located in the Grand Canyon, and northwest Phoenix.

County/Metro series

The fourteen maps of this series are created in batch with ArcObjects from templates, allowing each of them to have their own custom page layout. The county map here is of Yavapai County. The metro map shows the Tucson area and is used to help understand land management in federal and state lands surrounding this city.

Courtesy of the Arizona State Land Department.

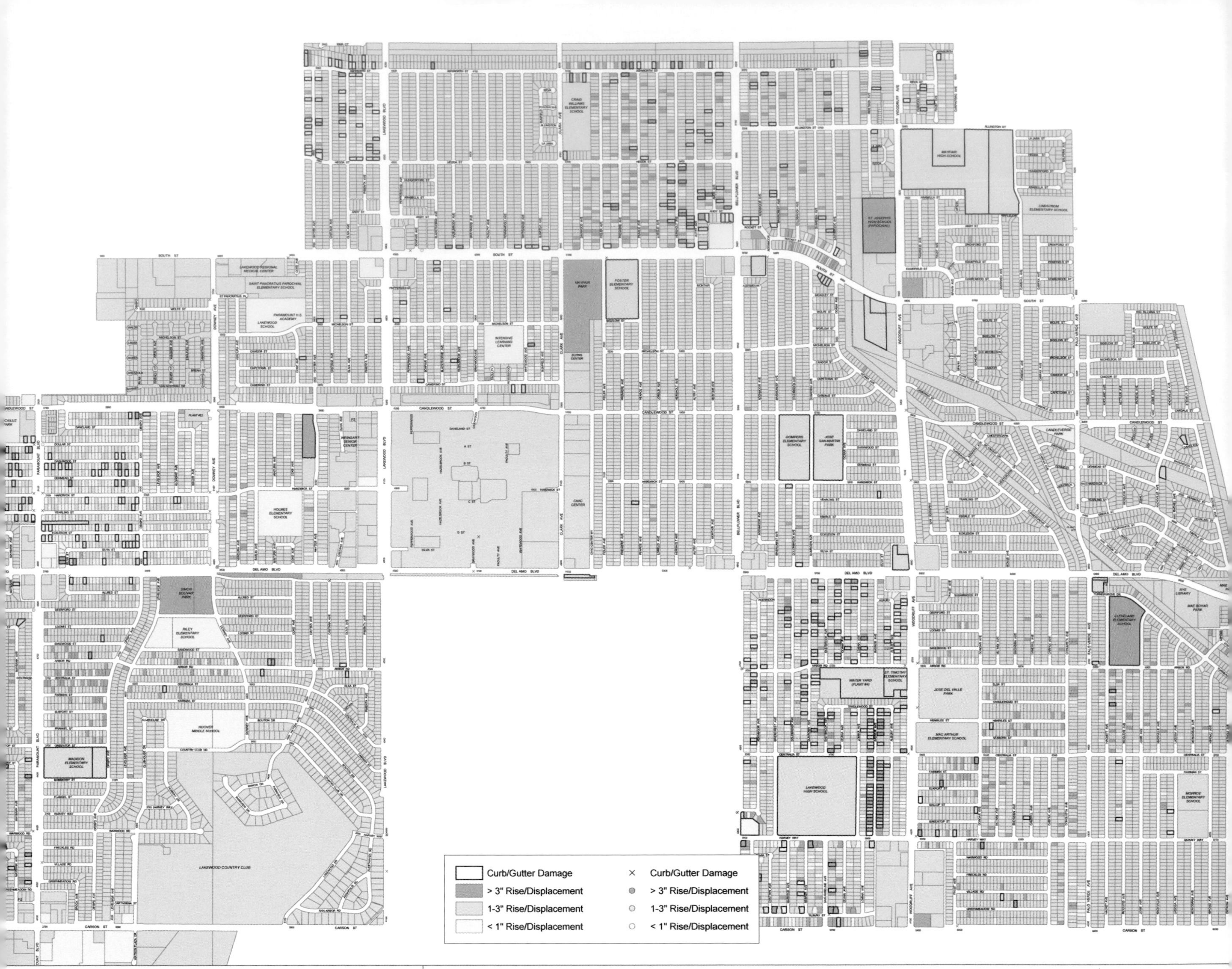

City of Lakewood
Lakewood, California, USA
By Michael Jenkins Jr.

Contact
Michael Jenkins Jr.
mjenkins@lakewoodcity.org

Software
ArcGIS Desktop 9.2

Printer
HP Designjet 5000ps

Data Sources
City of Lakewood, Los Angeles County Assessor

The City of Lakewood first implemented GIS in 1994. From the beginning, the city's Public Works Department recognized the power of GIS and quickly began to leverage its capabilities to help administer some of its programs. Hardscape management was a long-standing program that, until the mid-1990s, had "lived" only within a nonvisual, proprietary database. GIS gave Public Works staff the ability to visualize hardscape inventory data for the first time. That process revealed new patterns that empowered staff to make better-informed decisions. This map is an example of how GIS helps track hardscape damage within the city. Using this map, staff is able to monitor ongoing hardscape repair activities as well as plan future ones.

Unexpectedly, this map has also proven to be a valuable tool in the management of the city's urban forest. For years, the city planted only one tree species per city block. By overlaying the tree inventory GIS data on this map, staff was able to see a correlation between hardscape damage and certain tree species to easily target neighborhoods eligible for a replacement species.

Hardscape Damage Locations

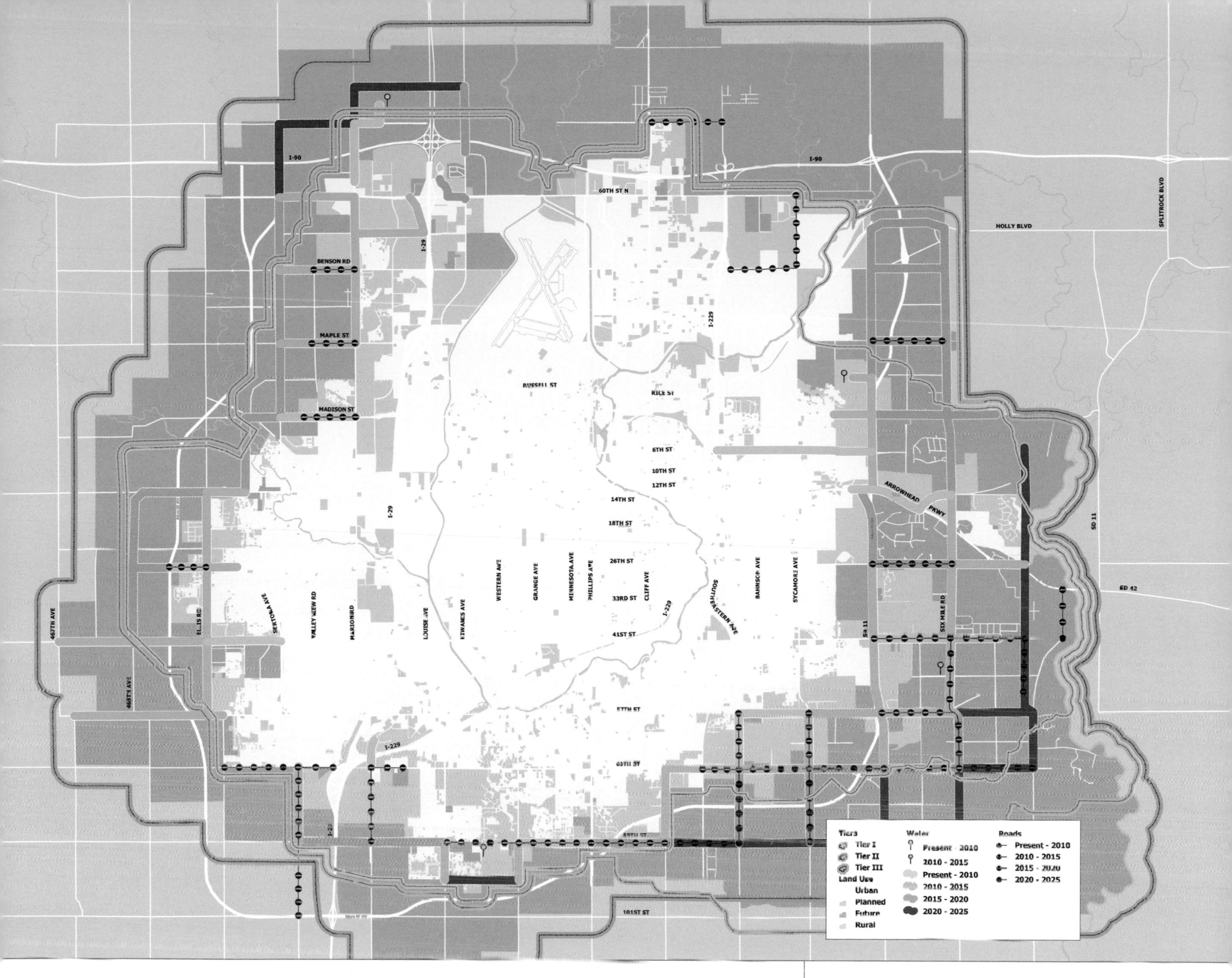

This map is part of a series that coalesces the City of Sioux Falls' growth management policies with its method of implementation. Since all new development is contingent on sanitary sewer availability, that is the primary basis for the city's tier structure. The maps are used throughout city government and by the general public. Data layers analyzed include numerous utilities, public facilities, and parks and recreation enhancements. Further, this map series visually demonstrates when the timing of like infrastructure, such as streets and water, are out of alignment.

Courtesy of Lauri B. Sohl, GISP, City of Sioux Falls.

City of Sioux Falls
Sioux Falls, South Dakota, USA
By Lauri B. Sohl, GISP

Contact
Lauri B. Sohl
lsohl@siouxfalls.org

Software
ArcGIS Desktop 9.2

Printer
Canon iPF8100

Data Source
City of Sioux Falls GIS

City of Sioux Falls Development Phasing Tiers

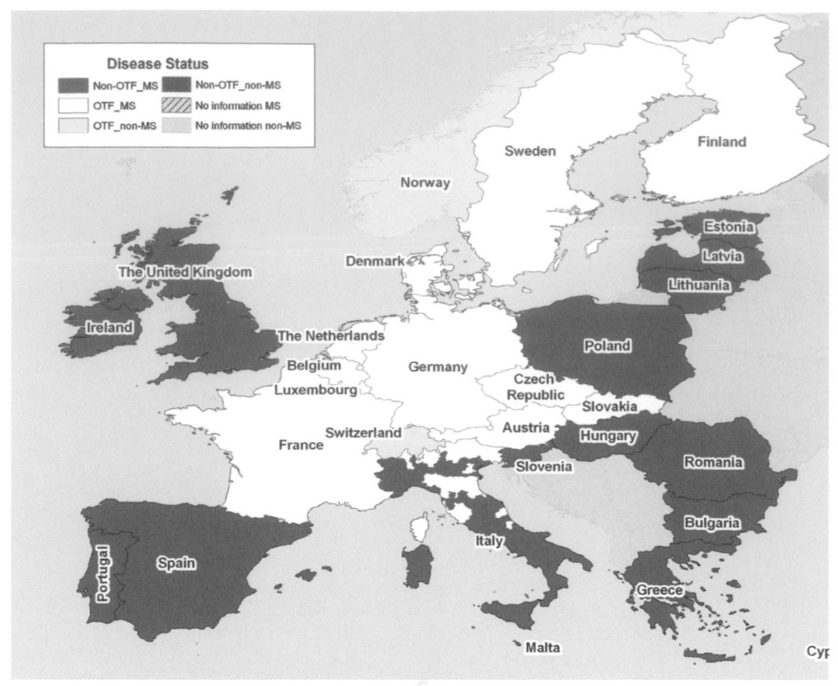

Status of Countries Regarding Freedom of Bovine Tuberculosis, 2007

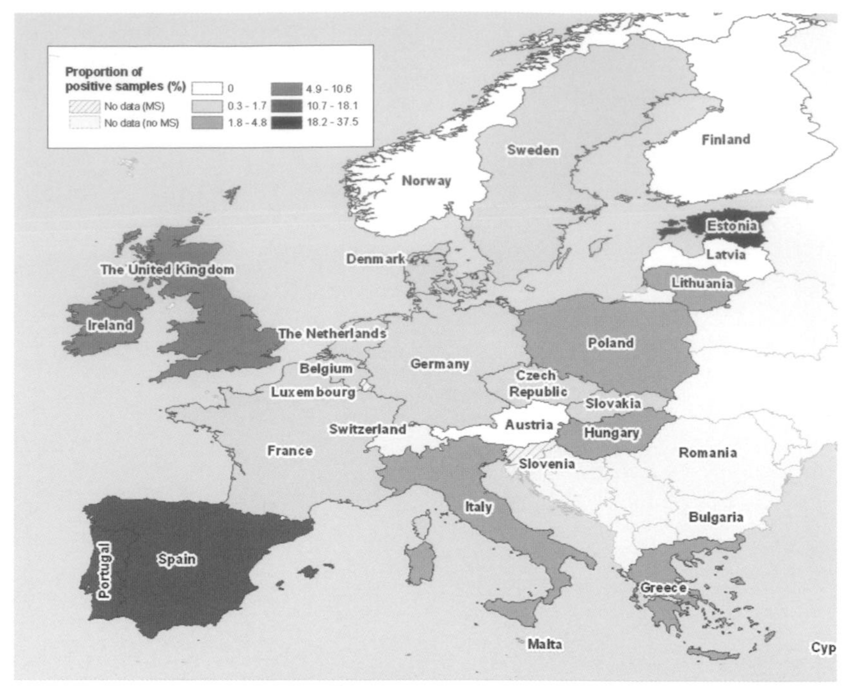

Proportion of Positive Samples to *Salmonella Spp.* in Breeding Gallus Gallus Flocks, 2006

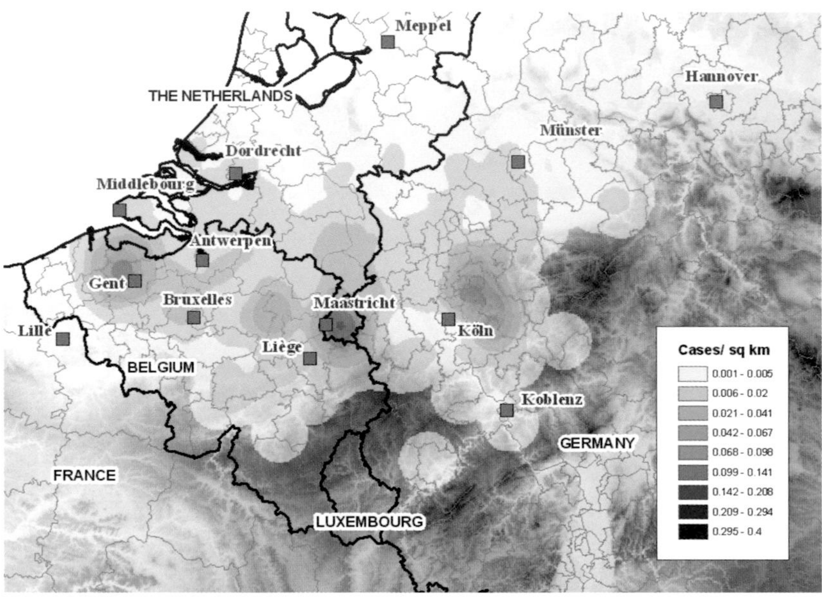

Kernel Smoothed Density Surface of Number of Cases for Small Ruminants Over Farm Density

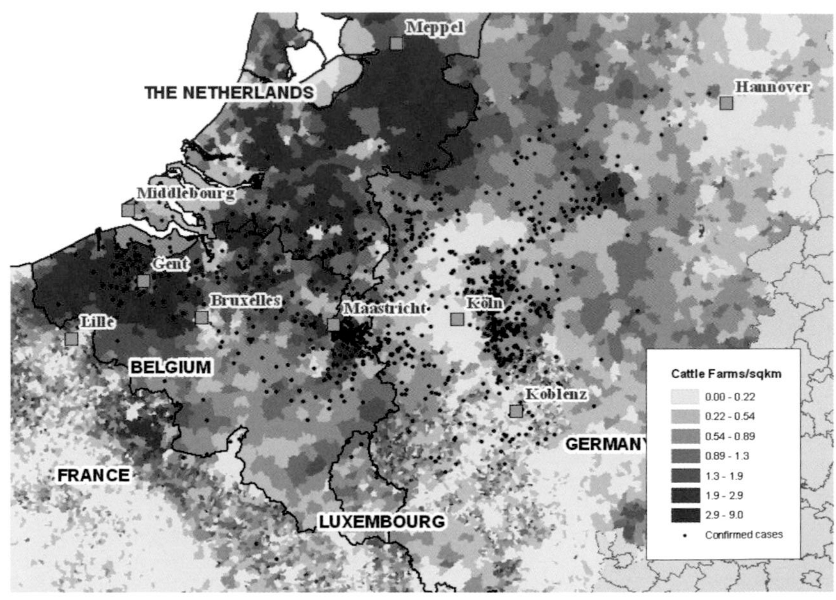

Distribution of Confirmed Btv8 Cases for Cattle Over Farm Density

European Food Safety Authority
Parma, Italy
By Francesca Riolo

Contact
Francesca Riolo
francesca.riolo@efsa.europa.eu

Software
ArcGIS Desktop 9.3, ArcGIS Server 9.3

Data Source
European Union Zoonoses Data Collection Database

Zoonoses are diseases that are transmissible from animals to humans directly or through ingestion of contaminated foodstuffs. Since 2004, according to the former Directive 92/117/EEC, European Union member states submit zoonoses data using an online reporting system and a central data repository developed and maintained by the European Food Safety Agency (EFSA).

Data collected covers over eleven zoonotic agents and zoonoses that include salmonella, verotoxin producing E. coli, tuberculosis due to Mycobacterium bovis, trichinella, and rabies. The data entered is compiled and analyzed in a yearly Community Summary Report on trends and sources of zoonoses. Maps showing the distribution of zoonoses are an integral part of this report. Further, with the advent of ArcGIS Server technology, and with the necessary confidentiality and data comparability precautions, the Web system offers a straightforward way to visualize and communicate the information, helping identify questions and hypotheses on the pattern and trends of diseases and their possible causes.

Besides the zoonoses data collection, EFSA investigates issues related to the origin and development of other epidemics, such as the Bluetongue virus in 2006. In this mandate, GIS tools and spatial analysis methods play a fundamental role.

Courtesy of Francesca Riolo, European Food Safety Authority.

A Web-based GIS for Zoonoses Monitoring

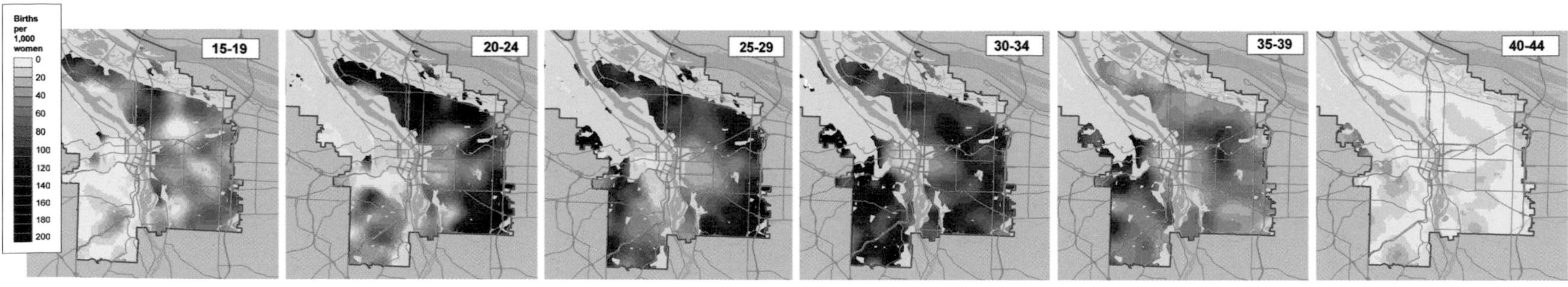

Map A: Age-Specific Fertility Rates for the Year 2000

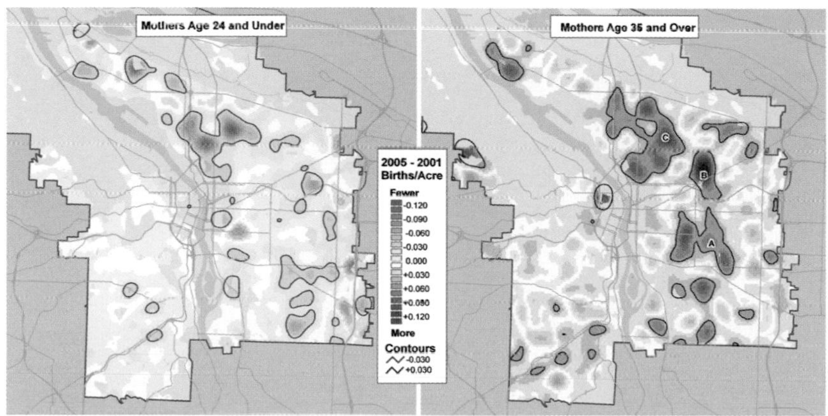

Map B: Change In Number of Births per Acre from 2001–2005, "Older Mother" Areas A, B, and C

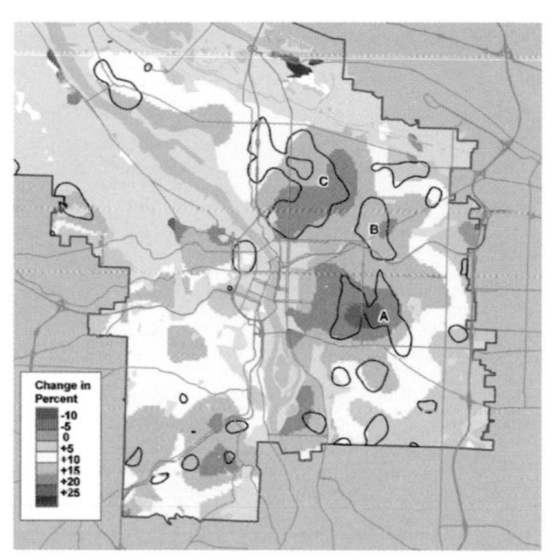

Map C: Change In Percent Persons Age 25 and Older with a 4-Year College Degree or Higher, 1990–2000

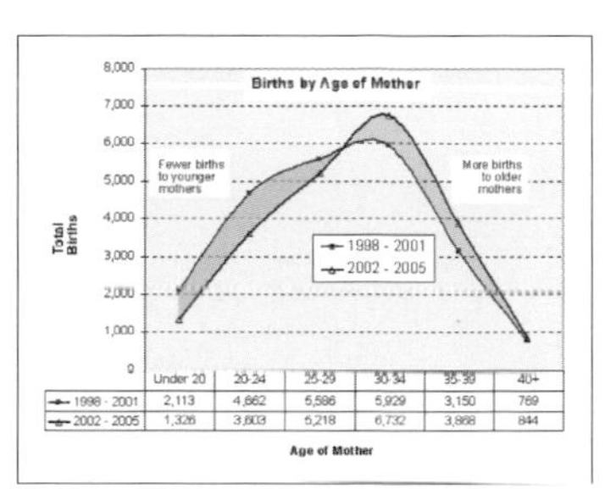

Chart 1: Births to Older Mothers Increased Recently Compared to Younger Mothers

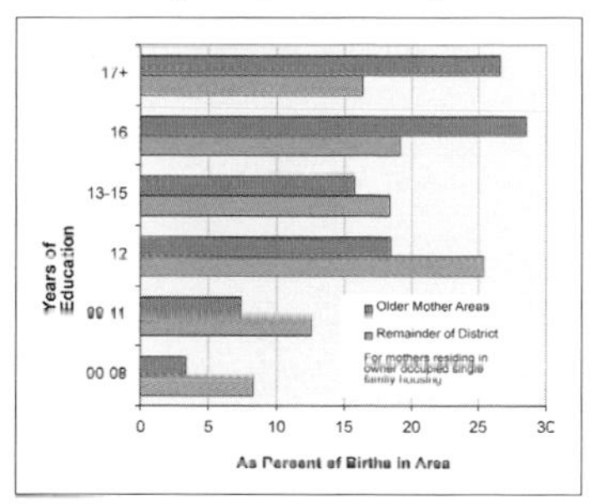

Chart 2: Years of Education by Area for Mothers in Single-Family Housing

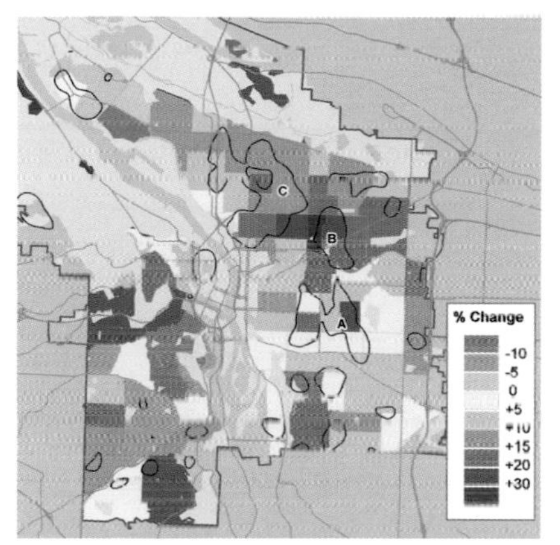

Map D: Cohort Trend for Households in Their 20s with Children Under Age 5, 1990–2000

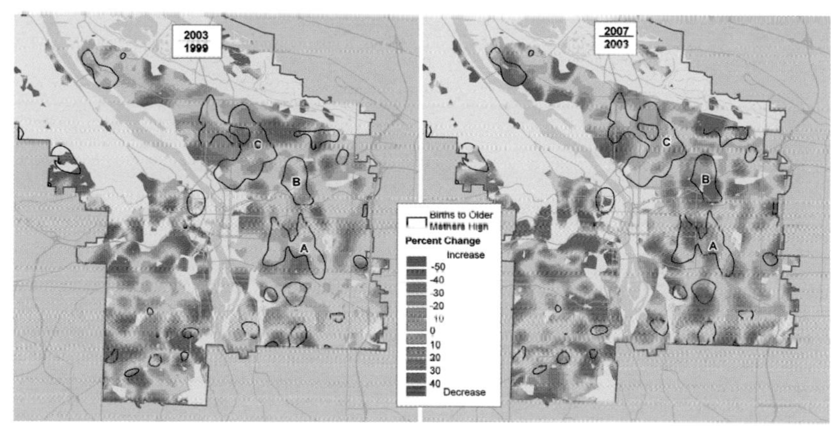

Map E: Cohort Trends for Grades K–2 Enrollment, 1999–2003 and 2003–2007

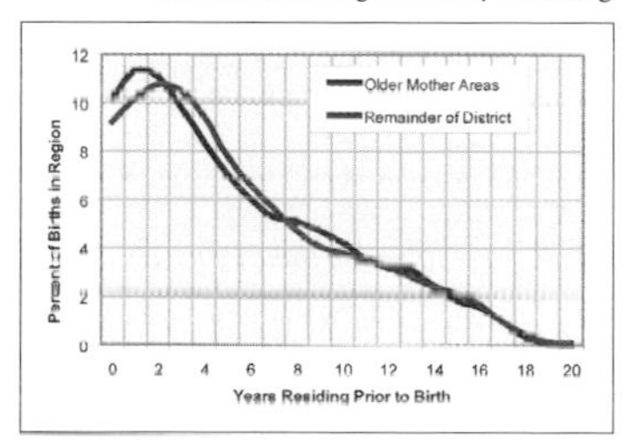

Chart 3: Years of Residence Prior to Birthing for Older Mothers in Single-Family Housing

These maps are part of an analysis of enrollment shifts in the Portland School District by the Population Research Center at Portland State University. They help explain how a rising number of births to older mothers has reversed the trend of declining school enrollments in several of the district's gentrifying neighborhoods.

Among the information shown on the maps are where the birth rates for older mothers are higher and areas where the density of births to older mothers increased substantially from 2001 to 2005. Maps and charts also show that these older mothers are a highly educated group, indentify gentrifying neighborhoods as measured by an increase in the percentage of highly educated persons, and show that the in-migration of these women began in the 1990s.

Courtesy of Portland State University.

Portland State University
Portland, Oregon, USA
By Richard Lycan and Charles Rynerson

Contact
Richard Lycan
lycand@pdx.edu

Software
ArcGIS Desktop 9.2, ArcGIS Spatial Analyst, Adobe Illustrator, Microsoft Excel

Printer
HP Designjet 500ps

Data Sources
Vital statistics, U.S. Census, Portland Metro Regional Land Information System, Portland Public Schools, U.S. Census of Population and Housing

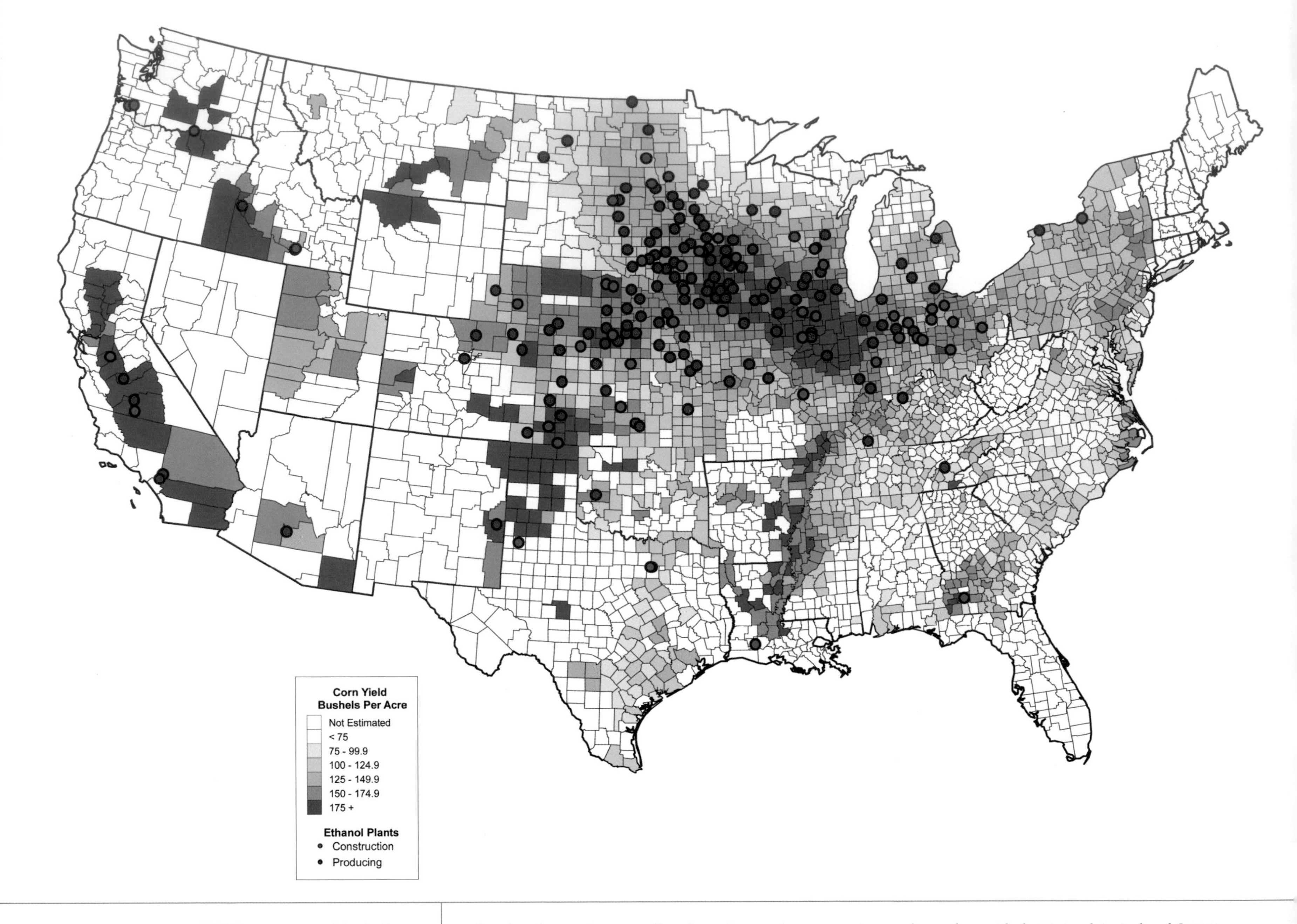

U.S. Department of Agriculture, National Agricultural Statistics Services
Fairfax, Virginia, USA
By Gail Wade

Contact
Gail Wade
gail_wade@nass.usda.gov

Software
ArcGIS Desktop 9.2

Printer
HP Designjet 5500

Data Sources
NASS County Estimates, Ethanol Producer Magazine

Ethanol producing plants as well as those plants under construction are shown along with the National Agricultural Statistics Services' (NASS) county estimates for corn for grain yield per harvested acre by county. Driven by growing ethanol demand, U.S. farmers planted over 15 percent more corn acres in 2007. The NASS Annual County Estimates Program provides for the collection of crop data through cooperative agreements with each state. NASS field offices set annual county estimates for crop acreage, yield, and production, and submit them to headquarters for official dissemination.

2007 Corn for Grain Yield per Harvested Acre by County and Ethanol Plants

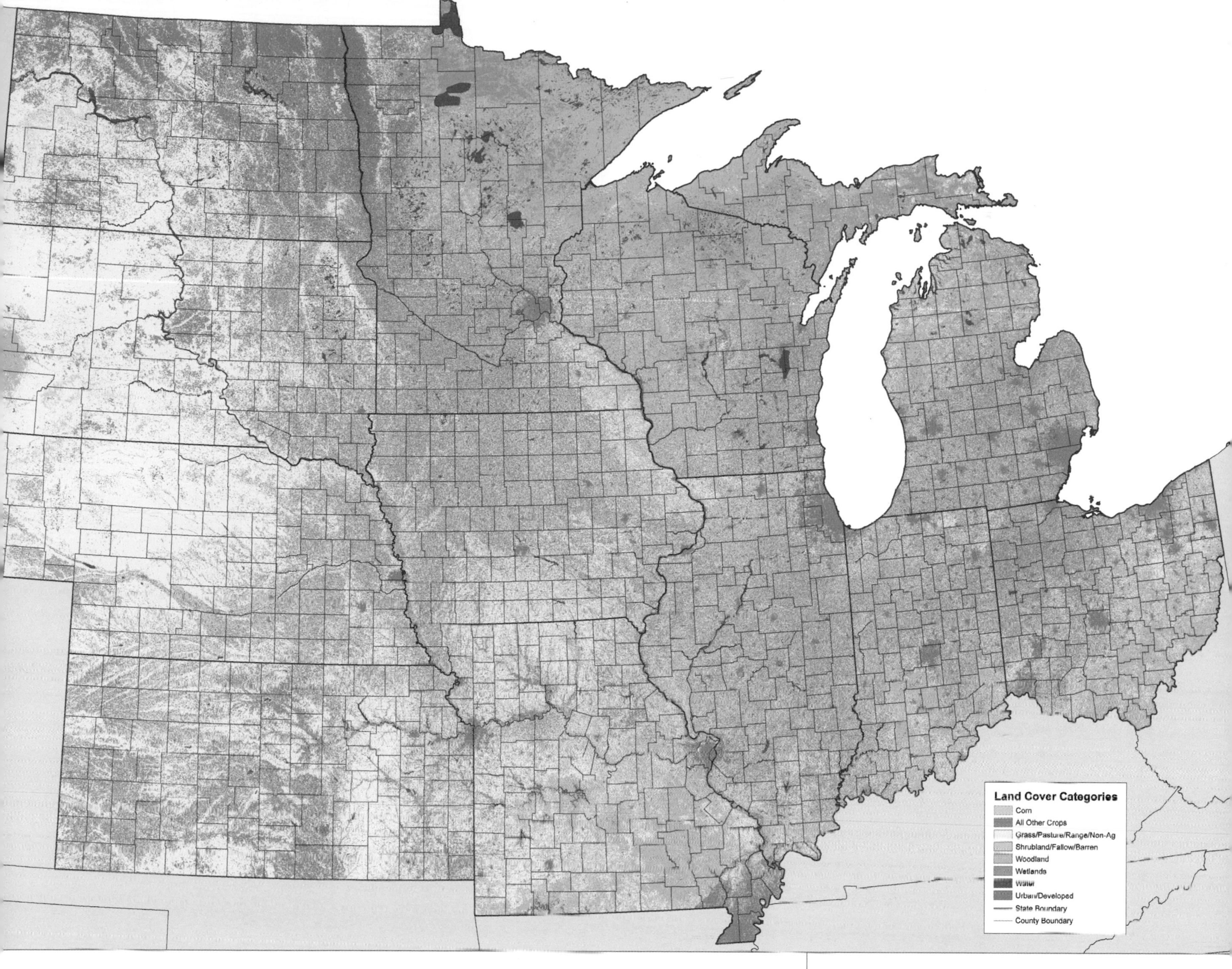

This map illustrates the predominance of corn grown throughout the midwestern United States in comparison to all other crops and noncrop land-cover categories. Harvested corn includes corn for grain/silage, sweet corn, popcorn, and ornamental corn.

The categorized Cropland Data Layer (CDL) imagery shown on the map was produced by the U.S. Department of Agriculture, National Agricultural Statistics Service (NASS). The CDL is used within NASS to generate supplemental acreage estimates of commodities for major agricultural states.

Courtesy of U.S. Department of Agriculture, National Agricultural Statistics Service.

U.S. Department of Agriculture, National Agricultural Statistics Service
Fairfax, Virginia, USA
By Claire Boryan, Mike Craig, Dave M. Johnson, Bob Seffrin, Patrick Willis, and Lee Ebinger

Contact
Rick Mueller
Rick_Mueller@nass.usda.gov

Software
ArcGIS Desktop 9.2, Rulequest See5, ERDAS Imagine 9.1

Printer
HP Designjet 5500 ps

Data Sources
Resourcesat-1 AWiFS, Terra MODIS, 2001 National Land Cover Dataset, Farm Service Agency Common Land Unit

Top 12 States for Harvested Corn Acreage—2007 Cropland Data Layers

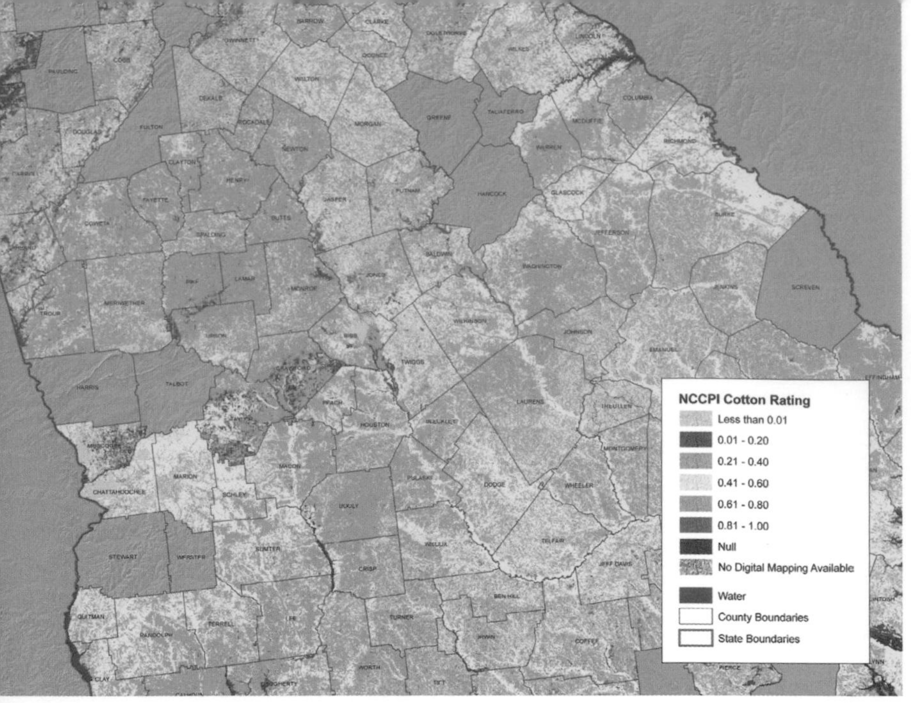

Georgia

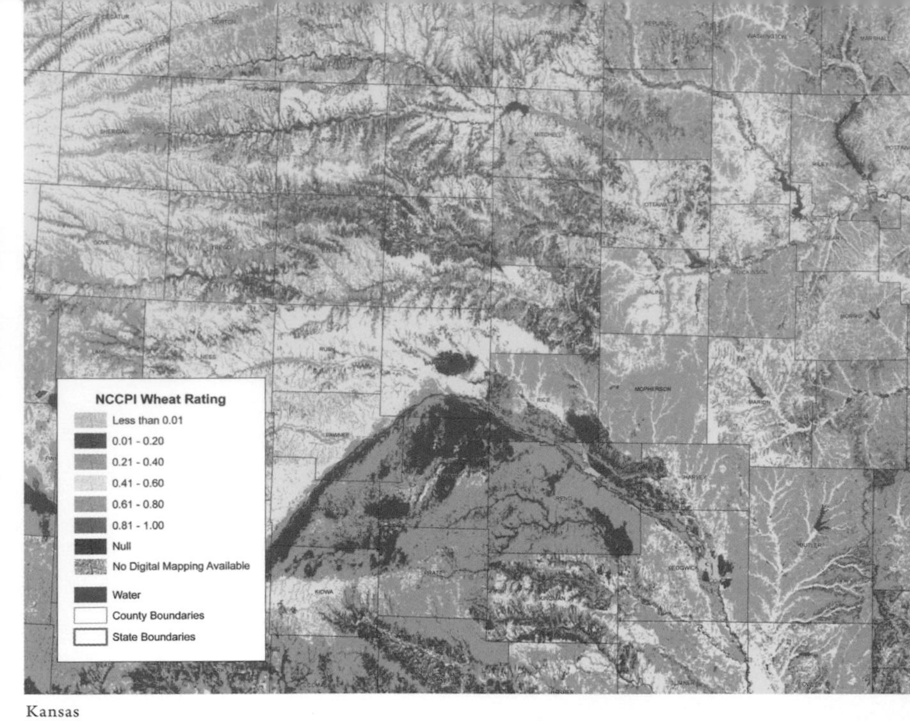

Kansas

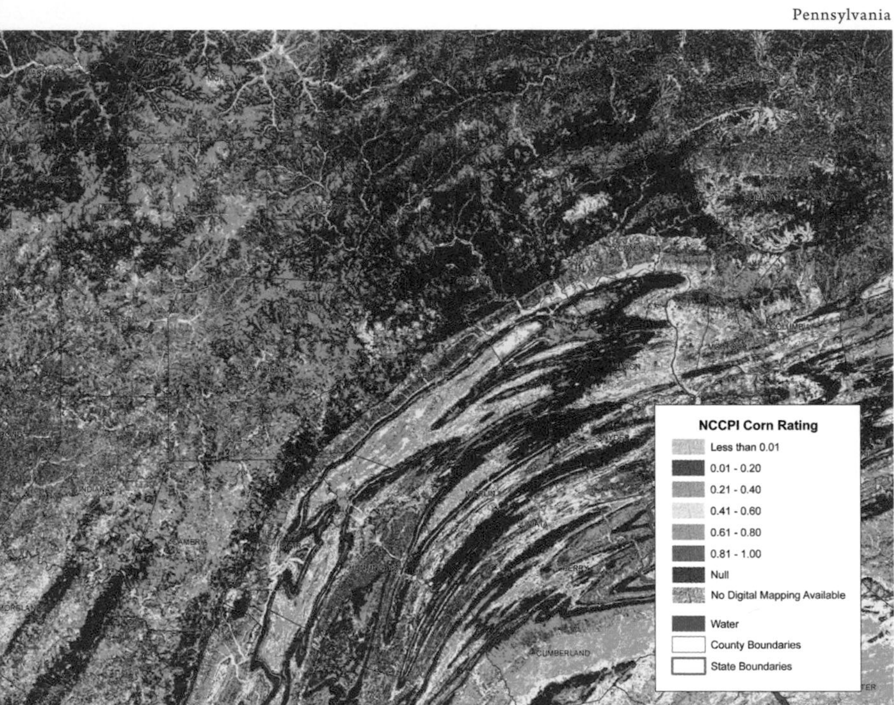

Pennsylvania

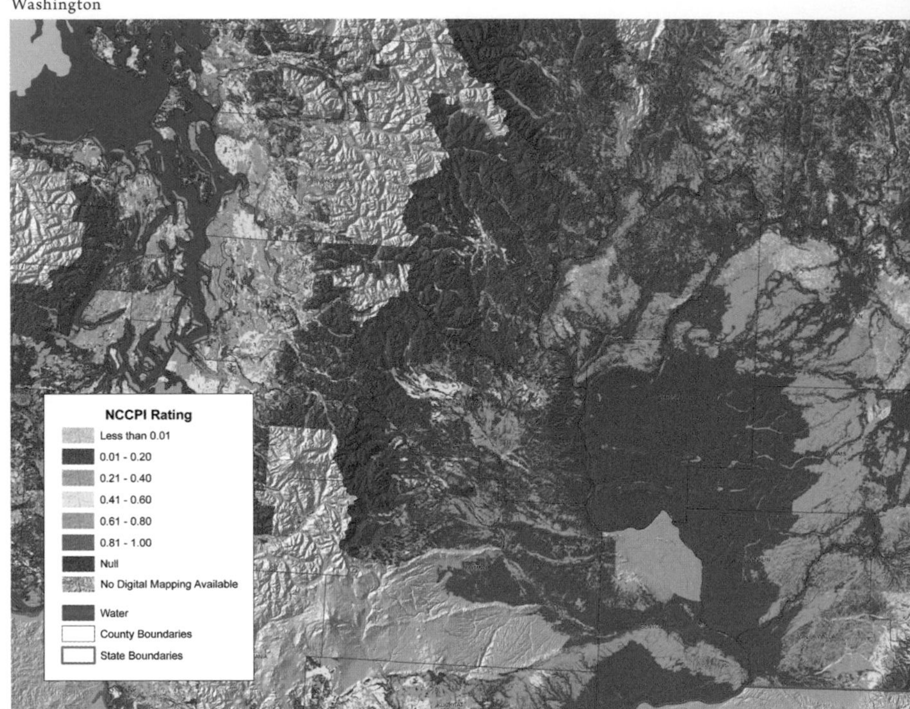

Washington

National Commodity Crop Productivity Index

U.S. Department of Agriculture, Natural Resources Conservation Service
Morgantown, West Virginia, USA
By Tim Prescott, Aaron Burkholder, Melissa Marinaro, Sharon W. Waltman, and Robert Dobos

Contact
Melissa Marinaro
melissa.marinaro@wv.usda.gov

Software
ArcGIS Desktop 9.2, Adobe Photoshop

Printer
HP Designjet 5500ps

Data Source
U.S. Department of Agriculture, Natural Resources Conservation Service

The National Commodity Crop Productivity Index (NCCPI) is a model that uses inherent soil properties, landscape features, and climatic characteristics to assign ratings for dry-land commodity crops such as wheat, cotton, sorghum, corn, soybeans, and barley. The model arrays Soil Survey Geographic Database map unit components from 0.01 to 1.0; components with the most desirable soil properties, landscape features, and climatic characteristics will display larger NCCPI values than soils with less desirable traits. The maps presented above are part of the Detailed Soil Survey Atlas, a national collection of state-centered maps prepared at a scale of 1:500,000 derived from U.S. Department of Agriculture soil geographic databases.

Courtesy of U.S. Department of Agriculture, Natural Resources Conservation Service.

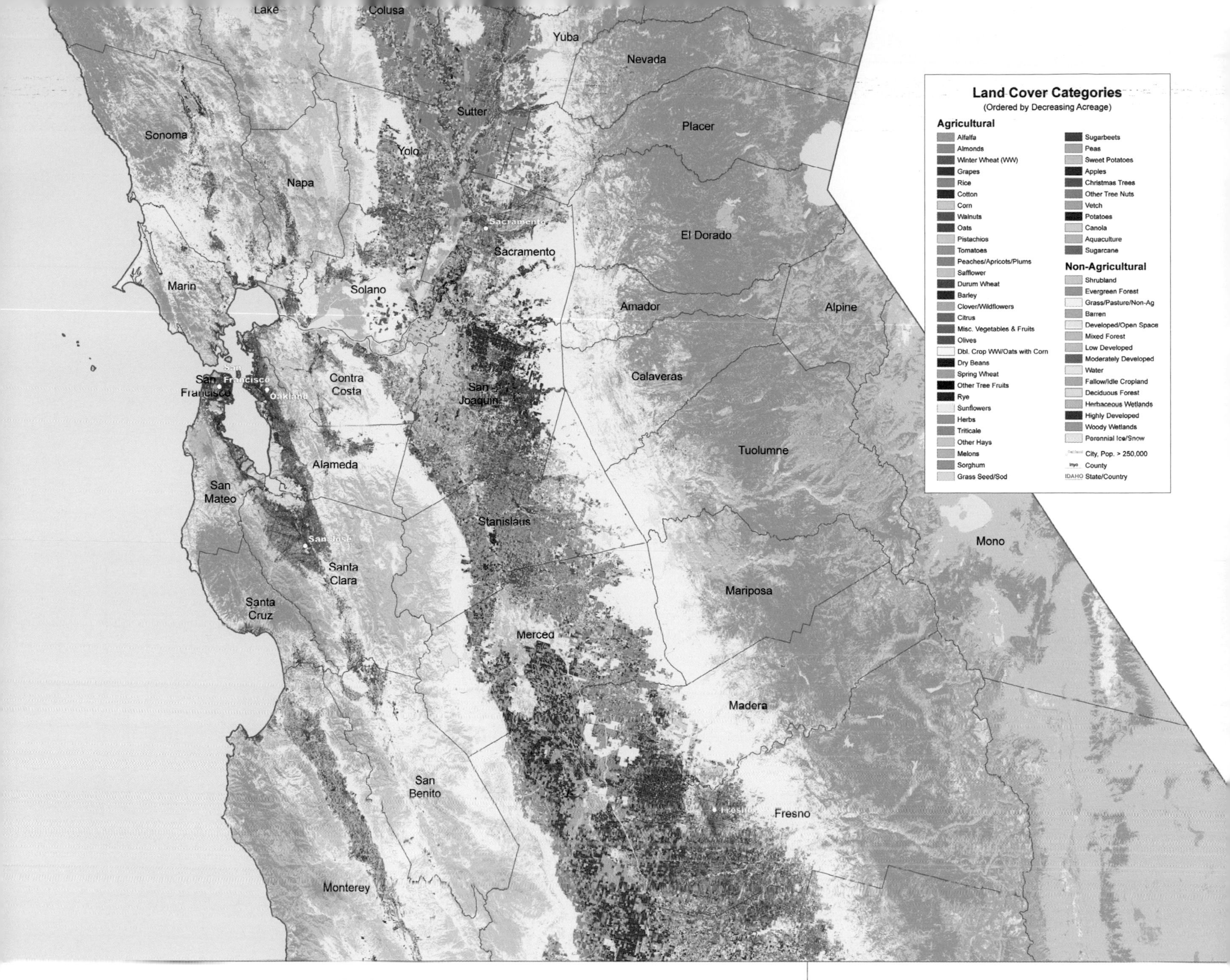

This map focuses on crop-specific land cover by identifying over forty crop categories and also includes major noncrop categories. The categorized Cropland Data Layer imagery shown on the map was produced by the National Agricultural Statistics Service (NASS) of the U.S. Department of Agriculture (USDA).

A decision-tree classification approach was applied using ground-truth data from NASS and the USDA Farm Service Agency; a combination of satellite imagery from Indian Remote Sensing Advanced Wide Field Sensor, and National Aeronautics and Space Administration Moderate Resolution Imaging Spectroradiometer sensors; and ancillary data sources.

Courtesy of U.S. Department of Agriculture, National Agricultural Statistics Service.

U.S. Department of Agriculture, National Agricultural Statistics Service
Fairfax, Virginia, USA
By Patrick Willis and Lee Ebinger

Contact
Rick Mueller
Rick_Mueller@nass.usda.gov

Software
ArcGIS Desktop 9.2, Rulequest See5, ERDAS Imagine 9.1

Printer
HP Designjet 5500ps

Data Sources
2001 National Land Cover Dataset, Terra Moderate Resolution Imaging Spectroradiometer, Farm Service Agency Common Land Unit, NASS June Area Survey data

California's Agricultural Land Cover—2007 Cropland

General Directorate of Forestry
Yenimahalle, Ankara, Turkey
By Umut Adigüzel and Metin Kocaeli

Contact
Umut Adigüzel
umutadiguzel@ogm.gov.tr

Software
ArcGIS Desktop 9.2

Printer
HP Designjet 800ps

Data Source
General Directorate of Forestry

This is a forest stand-type map produced for one of the 1,308 subdistricts of Turkey's forest management plan. These plans are renewed every ten years. Aerial photos are combined with the field work to create final forest maps. The database structure, standards, and symbols are developed for Turkey.

Courtesy of Umut Adigüzel, General Directorate of Forestry.

Beydagi Forest Map

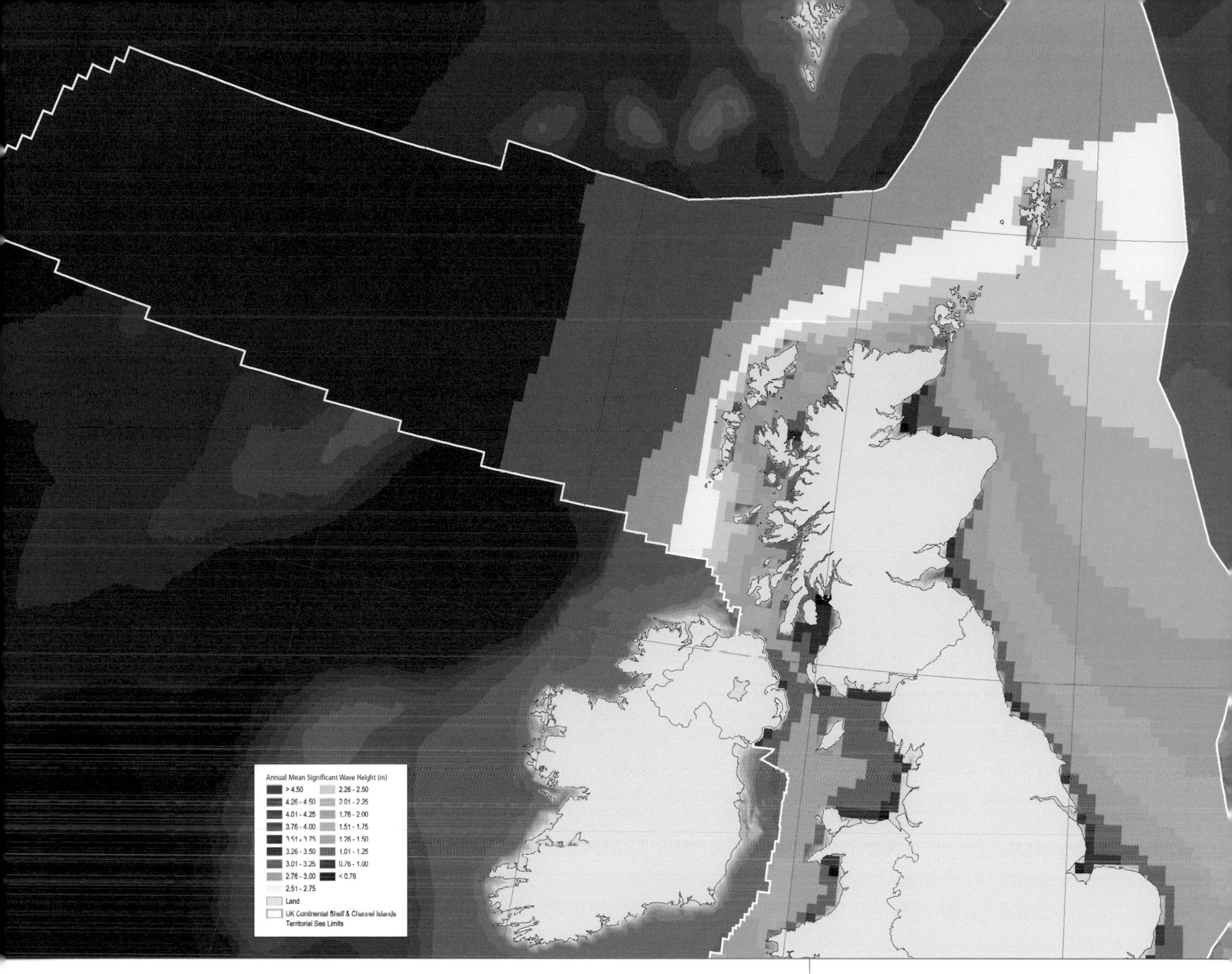

The Atlas of UK (United Kingdom) Marine Renewable Energy Resources stands as an example of an interactive data-management system, delivered across a range of GIS enabled formats: hard-copy maps, desktop application, and Web application. The range of delivery platforms has enabled the Renewables Atlas to become a highly regarded data resource outlining the potential opportunities within the UK marine renewable energy sector (tide, wind, and wave) for stakeholders and developers alike. The atlas now represents the most detailed regional description of potential marine energy resources in UK waters ever completed at a national scale, and is being used to help guide policy and planning decisions for future site leasing rounds.

ABP Marine Environmental Research Ltd.
Southampton, Hampshire, United Kingdom
By Chris Jackson, David Petrey, Colin Bell, and Andy Saulter

Contact
Chris Jackson
cjackson@abpmer.co.uk

Software
ArcGIS Desktop 9.2

Data Sources
Met Office and Proudman Oceanographic Laboratory

Atlas of UK Marine Renewable Energy Resources

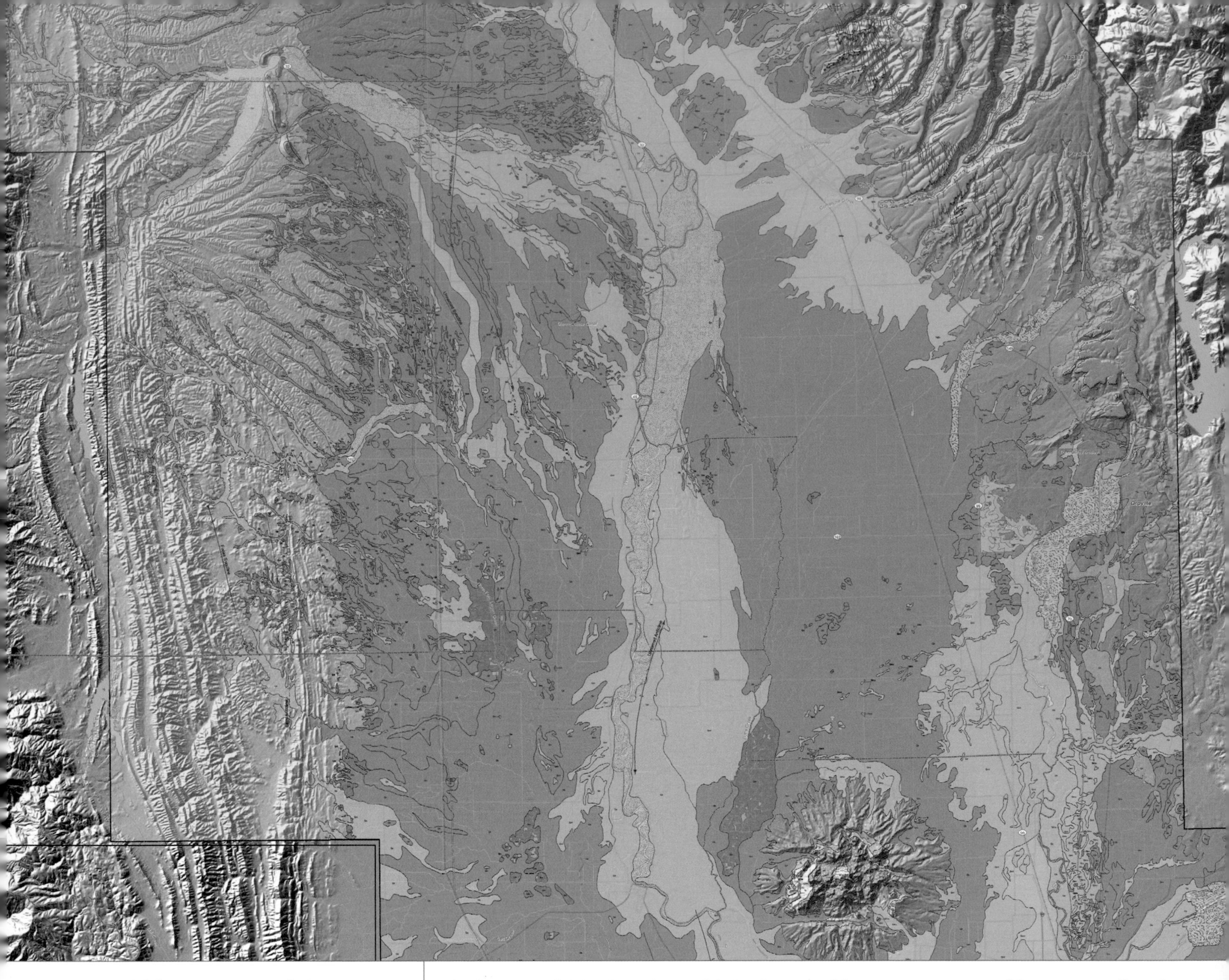

California Department of Water Resources
Red Bluff, California, USA
By Jonathan Mulder

Contact
Jonathan Mulder
mulder@water.ca.gov

Software
ArcGIS Desktop

Printer
HP Designjet 5500ps

Data Source
Geologic Map of the Late Cenozoic Deposits of the Sacramento Valley and Northern Sierran Foothills, California

This map is a modified digital reproduction of the "Geologic Map of the Late Cenozoic Deposits of the Sacramento Valley and Northern Sierran Foothills, California," by Edward J. Helley and David S. Harwood (USGS Publication MF-1790, 1985).

This map was created by scanning the five-sheet set of the original Helley and Harwood map, georeferencing the scanned images, and digitizing the lithologic contacts and other geologic information in AutoCAD 2006. The digitized map was then colored and symbolized in ArcGIS Desktop 9.0 software. The accuracy of the digitized lines is within the accuracy of the originally drafted lines on the paper copy. In general, the width of the contact lines on the paper copy extends to about 20 meters (66 feet).

Minor topological mistakes (such as identical rock units on both sides of a lithologic contact or unclosed polygons) and omissions (such as unidentified lithologic units) have been corrected to the best of the author's geologic expertise. Comparisons were made between the five-sheet set and the original Mylar and colored field sheets (as available) in addition to various geologic maps.

This map was prepared by Jonathan Mulder, engineering geologist, Department of Water Resources, Northern District, Geological Investigations Unit. Assistance with the geological interpretation was provided by Bruce Ross, engineering geologist. Assistance with the digitizing and map layout was provided by student assistants Casey Murray, Clint Andreasen, and Jeremiah Moody.

Courtesy of the California Department of Water Resources.

Geologic Map of the Late Cenozoic Deposits of the Sacramento Valley, California

Potential Aquifer Recharge Zones

Soil Map, Butte County

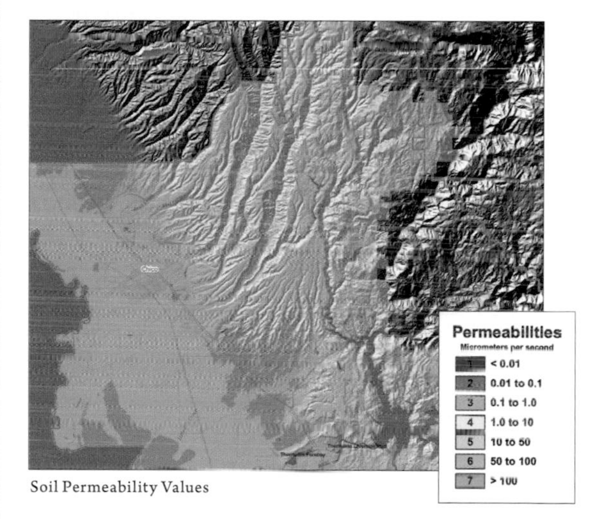

Soil Permeability Values

Some of the feature characteristics of potential aquifer recharge zones are soil permeability values and steepness of slope. In other words, water will tend to sink into the ground more readily in areas where the soil is more permeable and the slope is level or nearly level.

Soil permeability values and slope steepness values can be represented as layers in a GIS for further analysis. A new layer (Potential Aquifer Recharge Zones) can be created by combining the soil permeability values with the slope steepness values.

Courtesy of the California Department of Water Resources.

California Department of Water Resources
Red Bluff, California, USA
By Jonathan Mulder

Contact
Jonathan Mulder
mulder@water.ca.gov

Software
ArcGIS Desktop

Printer
HP Designjet 5500 ps

Data Sources
Natural Resources Conservation Service soil data and digital elevation model data

Determination of Potential Aquifer Recharge Zones Based on Soil Permeability and Slope, Butte County

Czech Geological Survey
Prague 1, Prague, Czech Republic
By P. Hanzl, Z. Krejci, K. Hrdlickova, L. Kondrova, and M. Klimova

Contact
Lucie Kondrova
lucie.kondrova@geology.cz

Software
ArcGIS Desktop 9.2

Printer
HP Designjet 5500ps

Data Source
Czech Geological Survey

Mongolia is a country of enormous mineral potential that attracts geologists and mining companies from all over the world. This creates a large demand for high-quality geological maps.

The Trans-Altai Gobi is a remote desert area of the southwestern wedge of Mongolia along the border with China. Part of the Trans-Altai Gobi belongs to the Great Gobi Protected Area, which was established in 1975 to protect a largely undisturbed part of the vast Gobi Desert, and to provide a refuge for the ancient terrestrial fauna of Central Asia. In 1991, the United Nations designated the Great Gobi as an international biosphere reserve.

One of the projects of a long-standing geological cooperation between Mongolia and the Czech Republic was oriented toward the geological mapping and geochemical reconnaissance of the Trans-Altai Gobi at a scale of 1:200,000. The map of the Trans-Altai Gobi at a scale of 1:500,000, published by Czech Geological Survey in 2008, is the summation of this work. All data was maintained in ESRI geodatabase format and cartographically processed in the ArcGIS environment.

Courtesy of Czech Geological Survey.

Geological Map of Trans-Altai Gobi

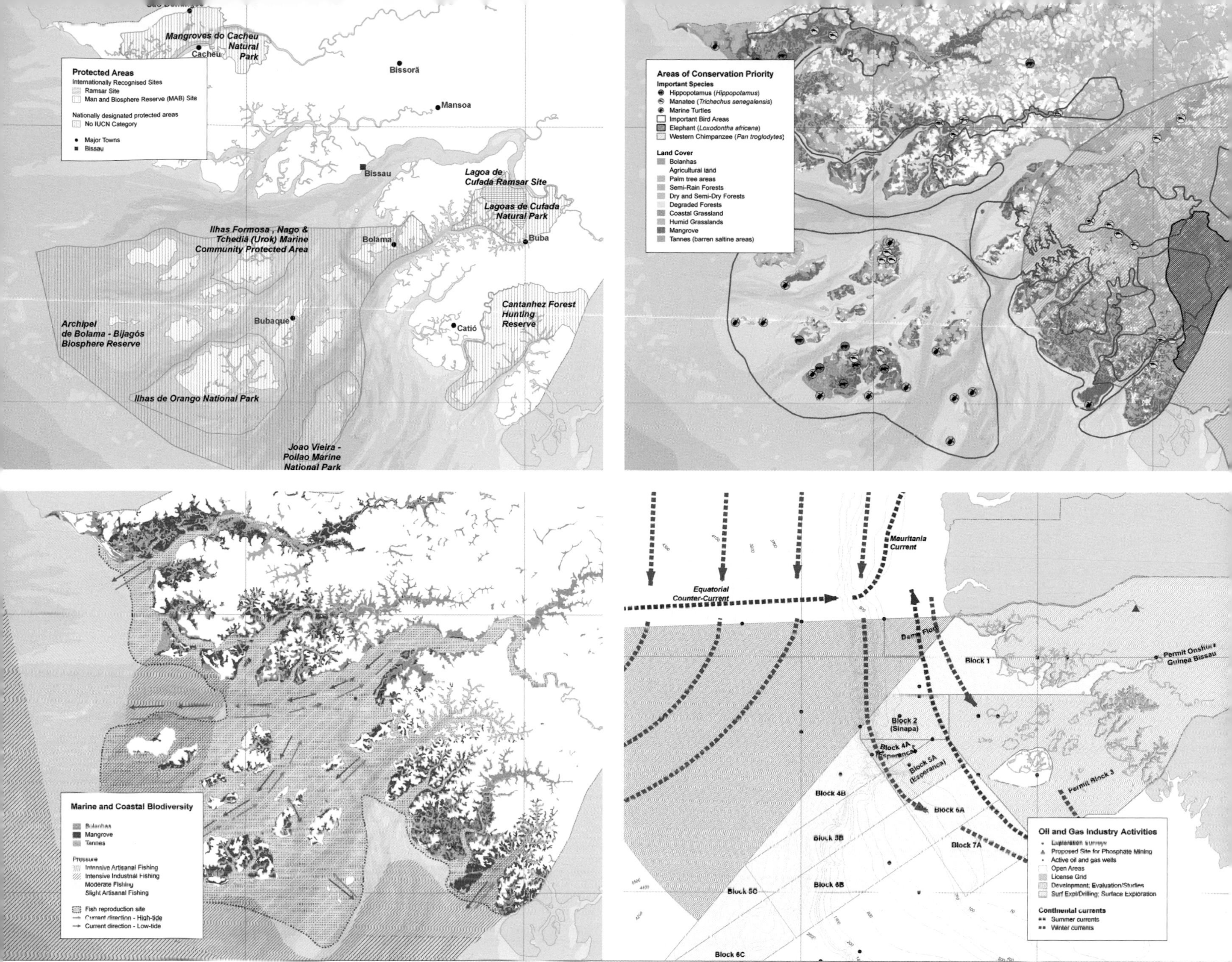

This map shows the effects of various activities on biodiversity. The United Nations Environment Programme—World Conservation Monitoring Centre (UNEP-WCMC) partnered with in-country experts from Guinea-Bissau to develop a synthesis map highlighting the potential pressures from oil and gas industry activities on the biodiversity of Guinea-Bissau. Such partnerships are leading to more accurate assessment of environmental pressures. A five-day workshop in Cambridge, United Kingdom, funded by World Wide Fund For Nature (WWF), resulted in combining UNEP-WCMC data with oil and gas exploration data from IHS Energy and data contributed by the visiting in-country experts. The combination of these data sources plus local knowledge and cartographic expertise at UNEP-WCMC has produced a powerful poster map that enables decision makers to incorporate key biodiversity information into their planning and development processes.

Data sources for this poster map included BISSA-SIG Database (GPC/INEP/UICN/IBAP/ Geomer Laboratory CNRS-Brest); IBAP: Managing Biodiversity for Secure Development, 2006; protected areas (WDPA): UNEP-WCMC, January 2007; J. Caldecott and L. Miles, 2005 (gorilla and chimpanzee data); GEBCO Digital Atlas bathymetry data, published by the British Oceanographic Data Centre on behalf of the International Oceanographic Commission (of UNESCO) and the International Hydrographic Organisation, 2003; and Petroleum Exploration & Production data: IHS, copyright 2007.

United Nations Environment Programme—World Conservation Monitoring Centre
Cambridge, Cambridgeshire, United Kingdom
By United Nations Environment Programme—World Conservation Monitoring Centre

Contact
Simon Blyth
Simon.Blyth@unep-wcmc.org

Software
ArcGIS Desktop 9.1

Printer
HP Designjet 1055cm

Data Sources
Various

Biodiversity and Perspectives on Oil, Gas, and Mining Exploitation in Guinea-Bissau

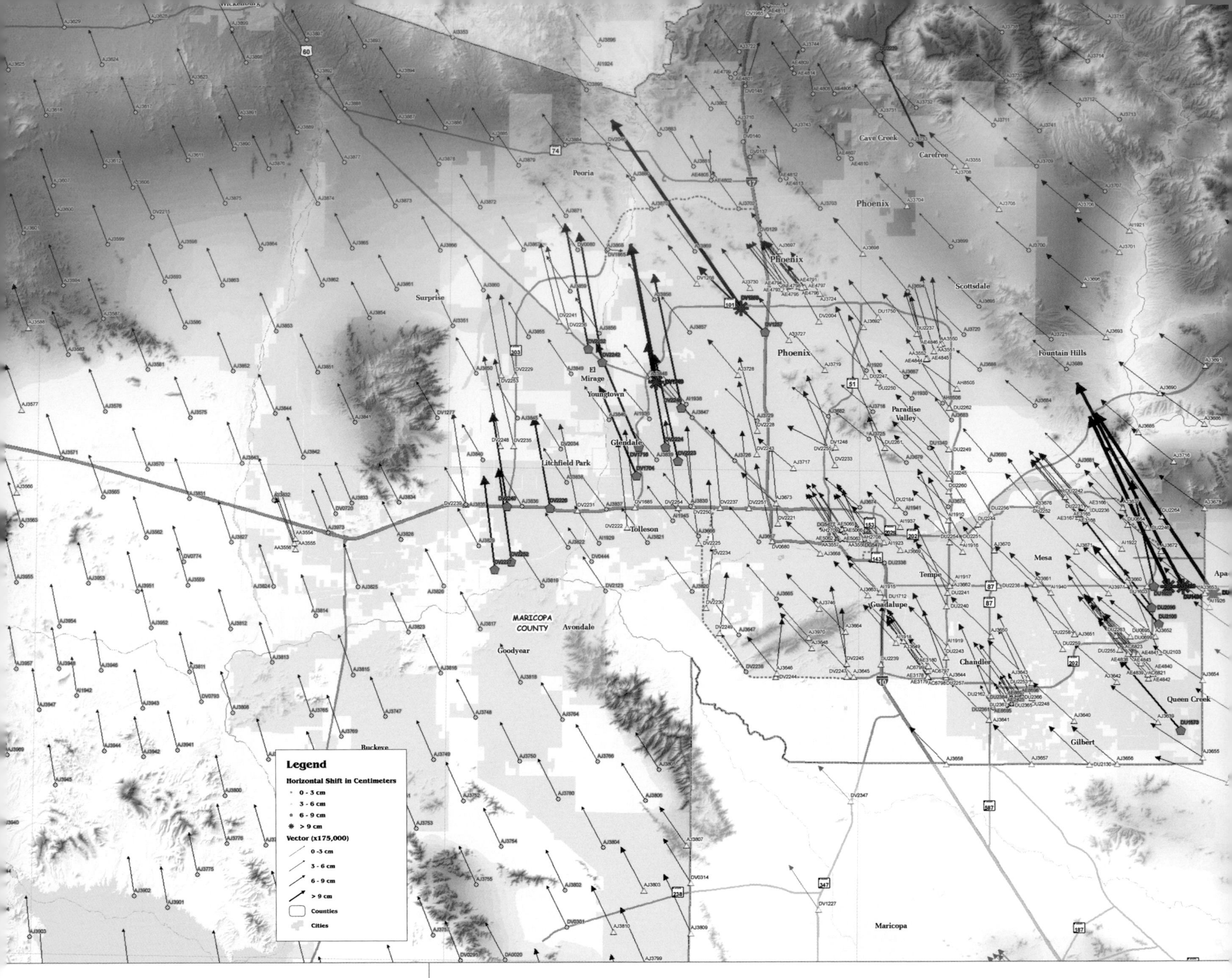

Arizona State Land Department

Phoenix, Arizona, USA
By Ronald Huettner

Contact

Ronald Huettner
rhuettner@land.az.gov

Software

ArcGIS Desktop 9.2

Printer

HP Designjet 4000ps

Data Sources

Tabular adjustment data, U.S. Geological Survey
10m data elevation model elevations

The example map shows the coordinate shifts for National Geodetic Survey control stations in Maricopa County, Arizona (the Phoenix metro area is in the central portion of the state), resulting from the national readjustment completed in February 2007. The coordinates being compared are for two realizations of North American Datum of 1983 (NAD83); NAD83 (1992) also known as the High Accuracy Reference Network (HARN) realization; and NAD83 (NSRS2007), also denoted as NAD83 (2007), for example on NGS datasheets. "NSRS" stands for National Spatial Reference System.

The issue being addressed here is of interest for those who have data layers with positional accuracies of a few centimeters with respect to NAD83 (HARN) and who wish to preserve the accuracy of those layers in the future.

Courtesy of the Arizona State Land Department.

2007 NSRS National Adjustment

This bedrock geology map of the Windsor-Wolfville area, Nova Scotia, was compiled at a scale of 1:50,000. ArcGIS software was used to digitize the map, design and populate the databases, and produce the cartographic product, including shaded relief illumination of the geology using a digital elevation model.

The map includes features such as bedrock units, anticlines, synclines, drill holes, faults, mineral occurrences, outcrops, shafts, trenches, quartz veins, structural data, glacial striations, quarries, and karst topography.

The area comprises a number of important geological terrains in Nova Scotia, including Triassic-Jurassic zeolite-bearing basalt of the North Mountain; Triassic sedimentary rock in the eastern end of the Annapolis Valley; carboniferous rocks of the Windsor Group, containing some of the largest gypsum quarries in the world; Cambro-Ordovician metasediments of the gold-producing Meguma Group; Devonian uranium- and tin-bearing granitic rocks of the South Mountain Batholith; and Devonian to Carboniferous zinc- and barite-bearing rocks of the Horton Group.

Acknowledgments: GIS databases and cartographic work by Angie L. Ehler, Jeff S. McKinnon, Brian E. Fisher, and other staff members of the Geoscience Information Services Section.

Courtesy of Nova Scotia Department of Natural Resources, Mineral Resources Branch.

Nova Scotia Department of Natural Resources
Halifax, Nova Scotia, Canada
By R. G. Moore, S. A. Ferguson, R. C. Boehner, and C. M. Kennedy

Contact
Brian E. Fisher
befisher@gov.ns.ca

Software
ArcGIS Desktop 9.3

Printer
HP Designjet 5500 ps

Data Source
Nova Scotia Department of Natural Resources, Mineral Resources Branch

Bedrock Geology Map of the Wolfville-Windsor Area

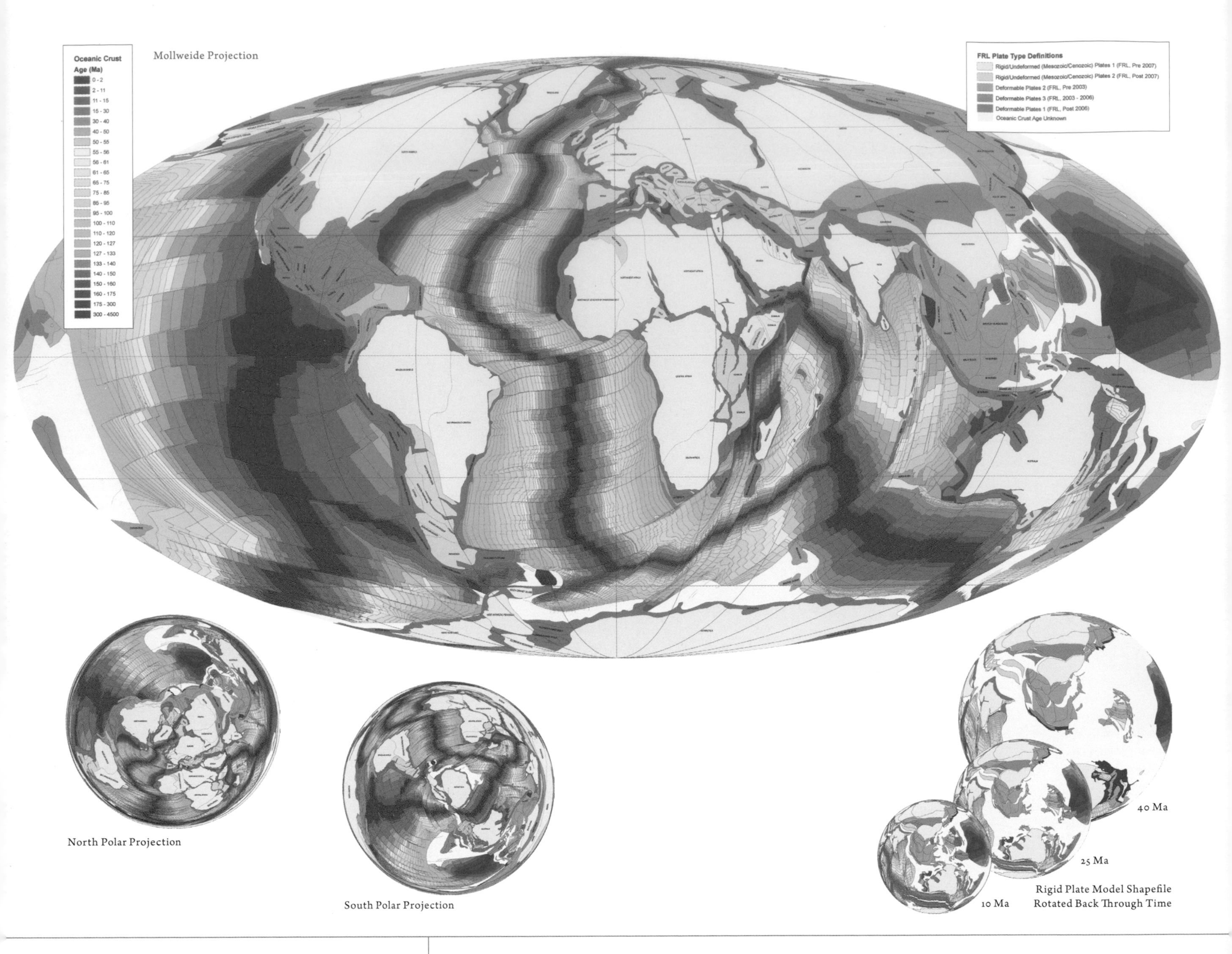

Fugro Robertson Limited

Llandudno, North Wales, United Kingdom

By Mike Goodrich, Lynne Hudson, Alberto Adriasola, Jim Harris, and Carl Watkins

Contact

Mike Goodrich

mjg@fugro-robertson.com

Software

ArcGIS Desktop 9.2

Printer

HP Designjet 5000ps

Data Sources

Primarily internal Fugro Robertson Limited (FRL) data, augmented by public domain literature, Shuttle Radar Topography Mission data elevation model data, FRL database of earthquake focal mechanisms, and gravity and magnetic data from Fugro Gravity and Magnetics

This map displays an overview of Fugro Robertson Limited's Plate Wizard project, which encompasses detailed global plate definitions, a dynamic model of plate reconstruction through geological time, a unique deformable plates methodology, geological control information, and a GIS front-end.

The project has as its starting point detailed global plate definitions, including defined rigid cores and deformable margins. These are based on the detailed regional plate models developed at Fugro Robertson Limited (FRL) over the last ten years, together with a comprehensive analysis of the near global passive margins and oceans gravity and magnetics dataset compiled by Fugro Gravity and Magnetics. This has been used in conjunction with FRL's global geological database to define a consistent global set of continent-ocean boundary definitions.

A key aspect of Plate Wizard is the development of a deformable-plates methodology for both convergent and divergent environments. Plate Wizard represents a major advance over the rigid plate models, with all their inherent problems, that have been available so far. The geological control information aspect of the project is feature linked in GIS to supporting databases, including geological control information and references. Finally, the GIS front-end allows full access to the plate polygons and rotation files, detailed browsing, access, reconstruction and deformation of both Plate Wizard and third-party data.

Plate Wizard

This map shows directional well surveys and how the engineering/survey data is converted to GIS format for display on a map. A map provides a two- or three-dimensional view of the directional survey, which aids in the geophysical visualization and interpretation that leads to the discovery of new oil and gas reservoirs. A map helps determine the volumetrics to calculate reserves and a map view of proposed drilling helps avoid intersecting an existing well.

This map of Plaquemines Parish, Louisiana, illustrates why directional surveys are important to oil and gas companies. A directional survey is a series of measured depths points down the well path with corresponding hole inclination and direction. From these points, the spatial position of the well path is calculated. The well directional survey information is given to the drilling company and reported to the state or federal regulatory agency, and operators.

The survey report becomes part of a proprietary database that is updated regularly. This information then becomes available to clients in various file formats suitable to their requirements. Well directional surveys have not been a highly publicized subject, but this is rapidly changing. Petroleum Place Energy Solutions, Inc. (P2ES) is the leading oil and gas data supplier creating well directional survey data standardized for mapping and three-dimensional interpretation purposes for the United States onshore. P2ES is the industry leader when it comes to "standardizing" for mapping purposes. This dataset is extremely valuable to the industry, because directional surveys (accurate to within plus-or-minus 4 feet) can now be easily mapped in a GIS environment.

Petroleum Place Energy Solutions, Inc./Tobin
San Antonio, Texas, USA
By Kandy Kenez

Contact

Kandy Kenez
kkenez@p2es.com

Software

ArcGIS Desktop 9.2, custom AMLs, custom proprietary Oracle DB, and Interface

Printer

Durst Lambda

Data Sources

Tobin aerial photography, Tobin SuperBase data products

Map Utilizing Well Directional Surveys

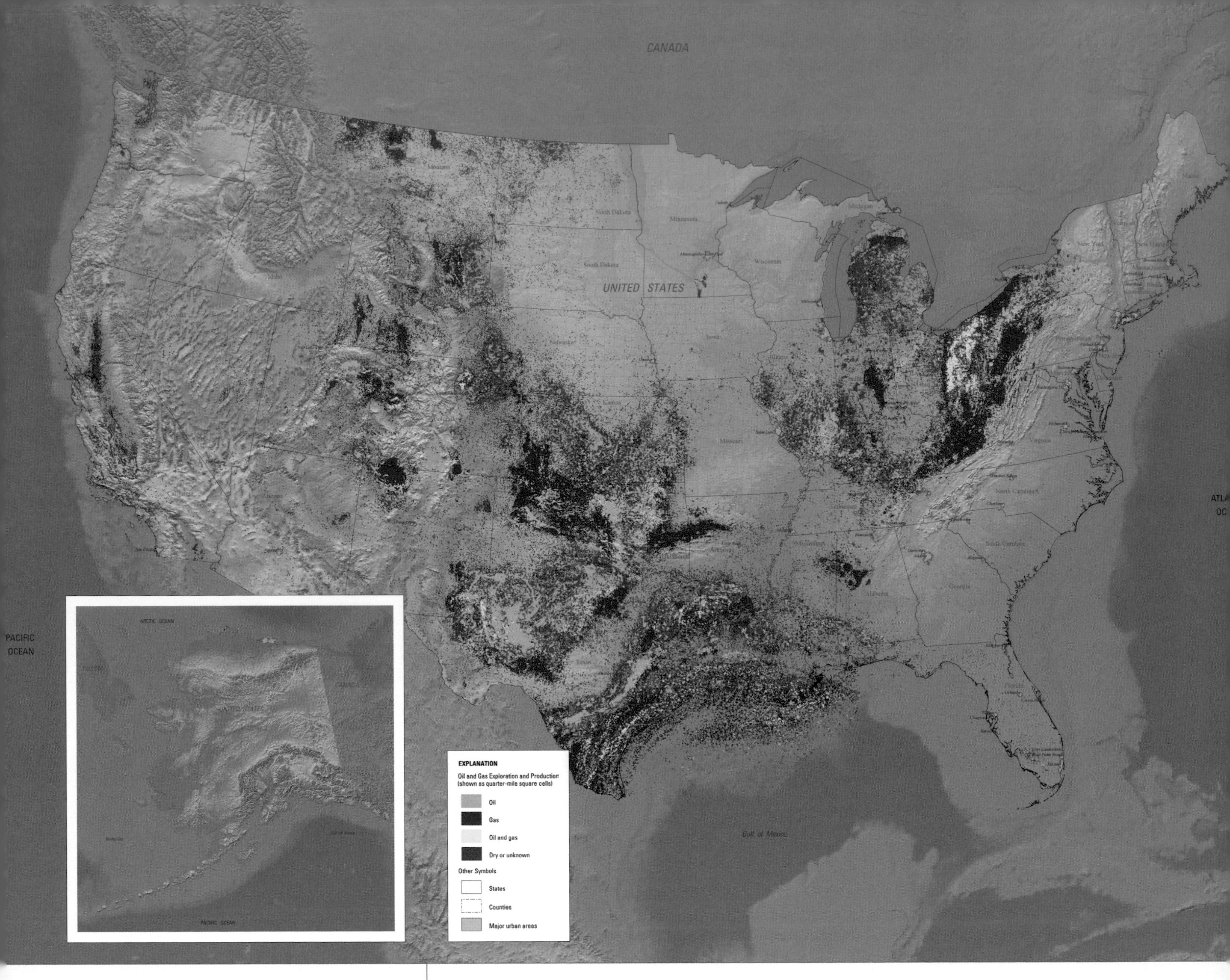

U.S. Geological Survey
Denver, Colorado, USA
By Laura R. H. Biewick

Contact

Laura R. H. Biewick
lbiewick@usgs.gov

Software

ArcGIS Desktop 9x, ArcIMS, Adobe Illustrator CS, Adobe Photoshop CS, and Adobe Acrobat 7.0

Printer

HP Designjet 5000ps

Data Sources

IHS Energy, Inc.; ESRI_Relief; U.S. Geological Survey National Map; Illinois, Indiana, Kentucky, Ohio, and Pennsylvania Geological Surveys; areas of historical oil and gas exploration and production in the United States from USGS Digital Data Series DDS-69-Q

This report contains maps and associated spatial data showing historical oil and gas exploration and production in the United States. Because of the proprietary nature of many oil and gas well databases, the United States was divided into one-quarter-square-mile cells, and the production status of all wells in a given cell was aggregated. Basemap reference data is included, using the U.S. Geological Survey (USGS) National Map, the National Atlas of the United States of America, the USGS and American Geological Institute Global GIS, and a world shaded relief map service from the ESRI Geography Network.

A hard-copy map was created to synthesize recorded exploration data from 1859, when the first oil well was drilled in the United States, to 2005. In addition to the hard-copy map product, the data has been refined and made more accessible through the use of GIS tools. The cell data is included in a GIS database constructed for spatial analysis via the USGS Internet map service or by importing the data into GIS software such as ArcGIS.

The USGS Internet map service provides a number of useful and sophisticated geoprocessing and cartographic functions via an Internet browser. Also included is a video clip of U.S. oil and gas exploration and production through time. This is a USGS Digital Data Series product that is available on the Web and on CD.

Areas of Historical Oil and Gas Exploration and Production in the United States

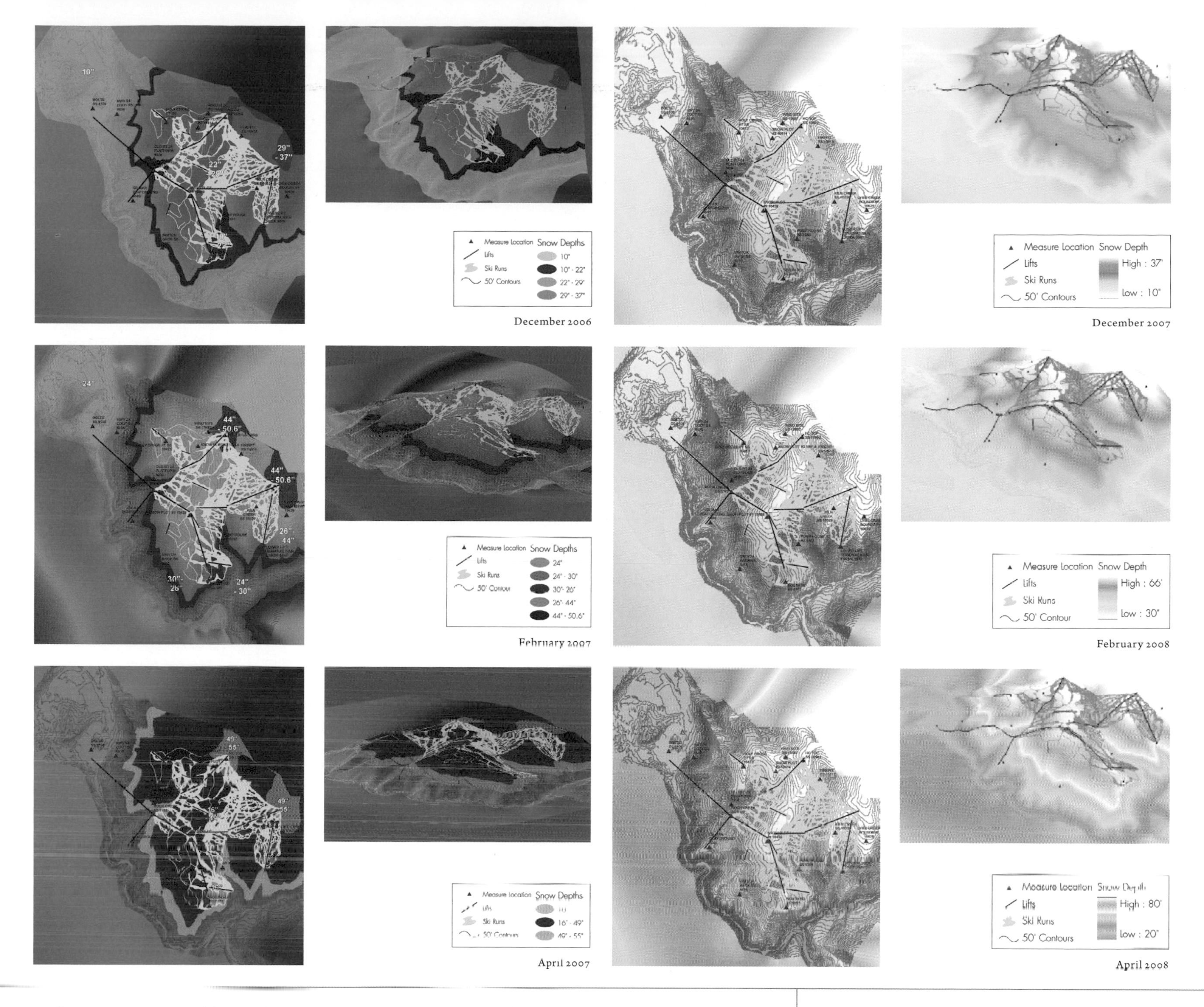

December 2006

December 2007

February 2007

February 2008

April 2007

April 2008

Battle Mountain is a proposed ski resort located in Colorado. By using lidar data with Merrick Advanced Remote Sensing (MARS) software, a highly accurate digital terrain model was created. Throughout the ski seasons of 2006–2007 and 2007–2008, snow depths were measured at set locations and regular intervals across the area. This data was compiled on a monthly basis and mapped using ArcGIS Desktop to create the different snow depth areas.

ArcScene is used to view the data in 3D to better understand the terrain related to the depth. These snow depths are used in the resort planning process for ski run locations, snowmaking needs, and possible development sites. Having these depths also allows for spring runoff calculations for the different basins on the development.

North Line GIS, LLC
Breckenridge, Colorado, USA
By Trip McLaughlin

Contact
Trip McLaughlin
maps@northlinegis.com

Software
ArcGIS Desktop 9.3, ArcScene, MARS

Printer
HP Designjet 5500ps

Data Sources
Dan Moroz Consulting and North Line GIS

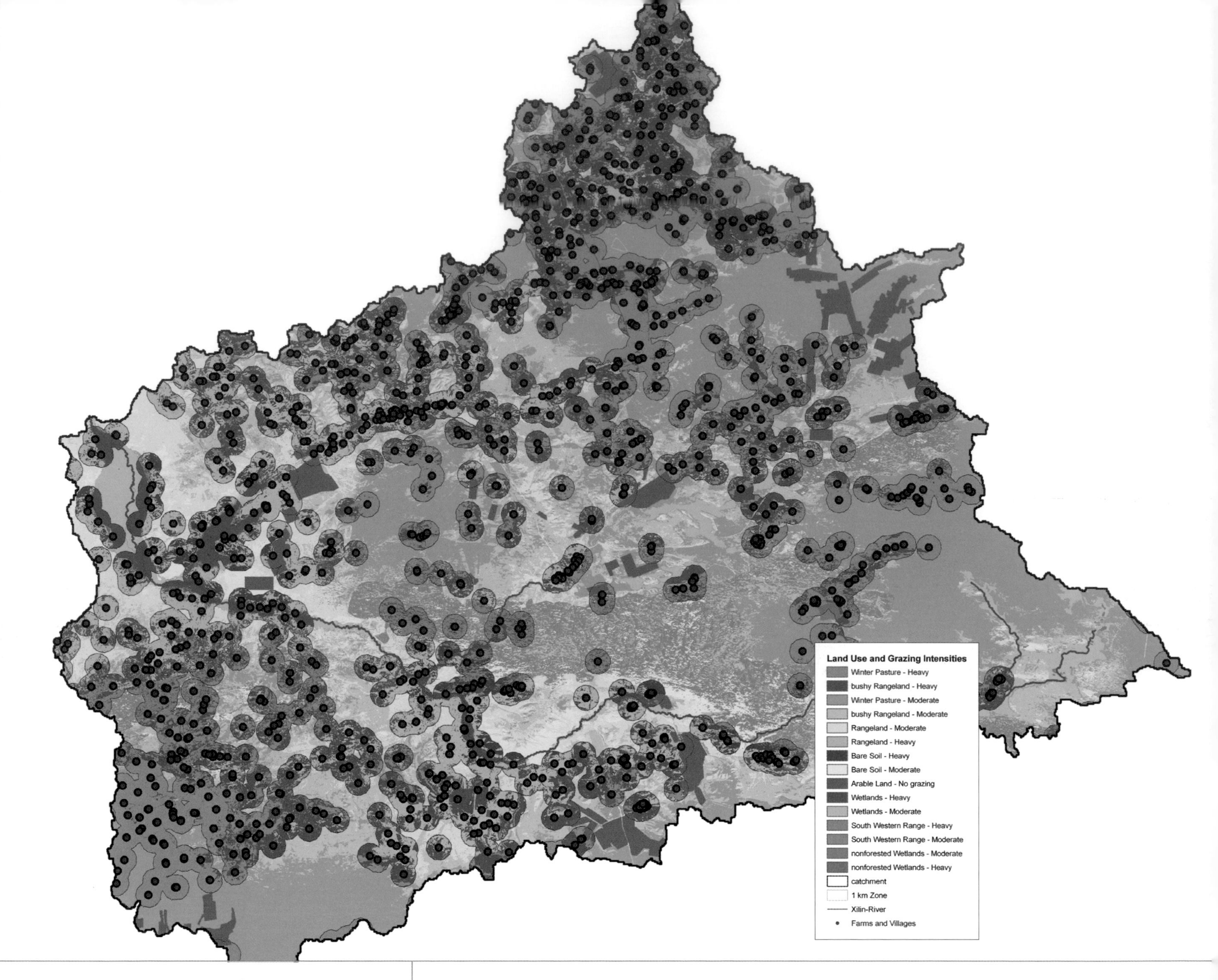

Justus-Liebig-Universität Giessen
Kirchhain, Hessen, Germany
By Johanna Schäfer

Contact
Johanna Schäfer
johanna-schaefer@gmx.net

Software
ArcGIS Desktop 9.2, ArcSWAT

Data Source
The MAGIM project

Asian, and especially Chinese, grassland ecosystems are threatened by desertification since population and stocking rates have been growing rapidly in recent decades. This study intended to find an appropriate land use within the Xilin River watershed in Inner Mongolia, China, to sustain the ecological balance and secure peoples' lives.

Normative land-use scenarios were created to predict the influence of different land-use changes in the water cycle in the Xilin River basin. One scenario assumed that environmental protection will increase and another assumed that agricultural production will rise to a maximum. Scenarios were analyzed with the SWAT model (Soil and Water Assessment Tool, U.S. Department of Agriculture) and ArcGIS Desktop 9.2.

The map shows the current land use of the Xilin river basin under different grazing intensities. As heaviest degradation is occurring around farms and villages, different grazing intensities were delineated by buffering rural settlements (1 km buffer threshold). In the next step, this map was joined with the current land-use map derived from Landsat 5 to produce the final land-use distribution. The current land use of the watershed was modeled with SWAT. The results were used to compare results from the other scenarios. The map was generated from data compiled by the MAGIM project (matter fluxes in grasslands of Inner Mongolia as influenced by stocking rate) funded by the German Research Foundation.

Courtesy of Johanna Schäfer.

Influences of Land-Use Change on Water Fluxes in the Xilin River Catchment, Inner Mongolia

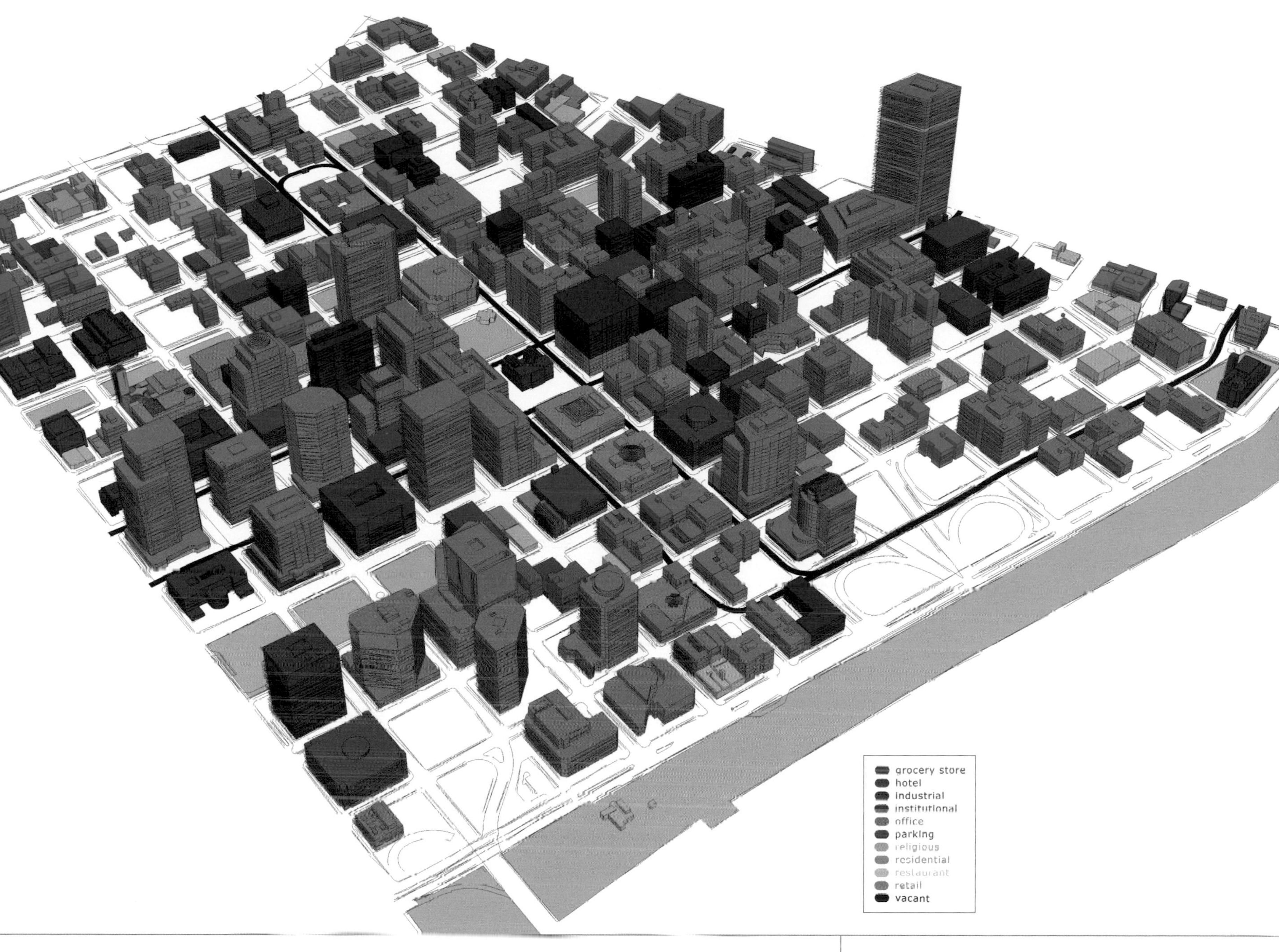

The City of Portland Bureau of Planning has developed a "3D land-use inventory" analytical GIS model that estimates the square footage of land use by building floor in Portland's central city. The model, developed in ArcInfo Workstation using ARC Macro Language (AML), produces a single GIS dataset that is used to visualize the land-use information in 3D (in Google SketchUp, ArcScene, or Google Earth).

The results are also used to generate total square footage statistics by land use for any geographic area within the central city. The two main inputs to the model are a database containing the percentage of land use by individual building floor, and a GIS-based 3D model of the city, both of which were created and are maintained by the Bureau of Planning.

Courtesy of City of Portland Bureau of Planning.

City of Portland Bureau of Planning
Portland, Oregon, USA
By Kevin Martin

Contact
Kevin Martin
kmartin@ci.portland.or.us

Software
ArcGIS Desktop 9.2, Adobe Illustrator

Printer
HP Designjet 1050 c

Data Source
City of Portland

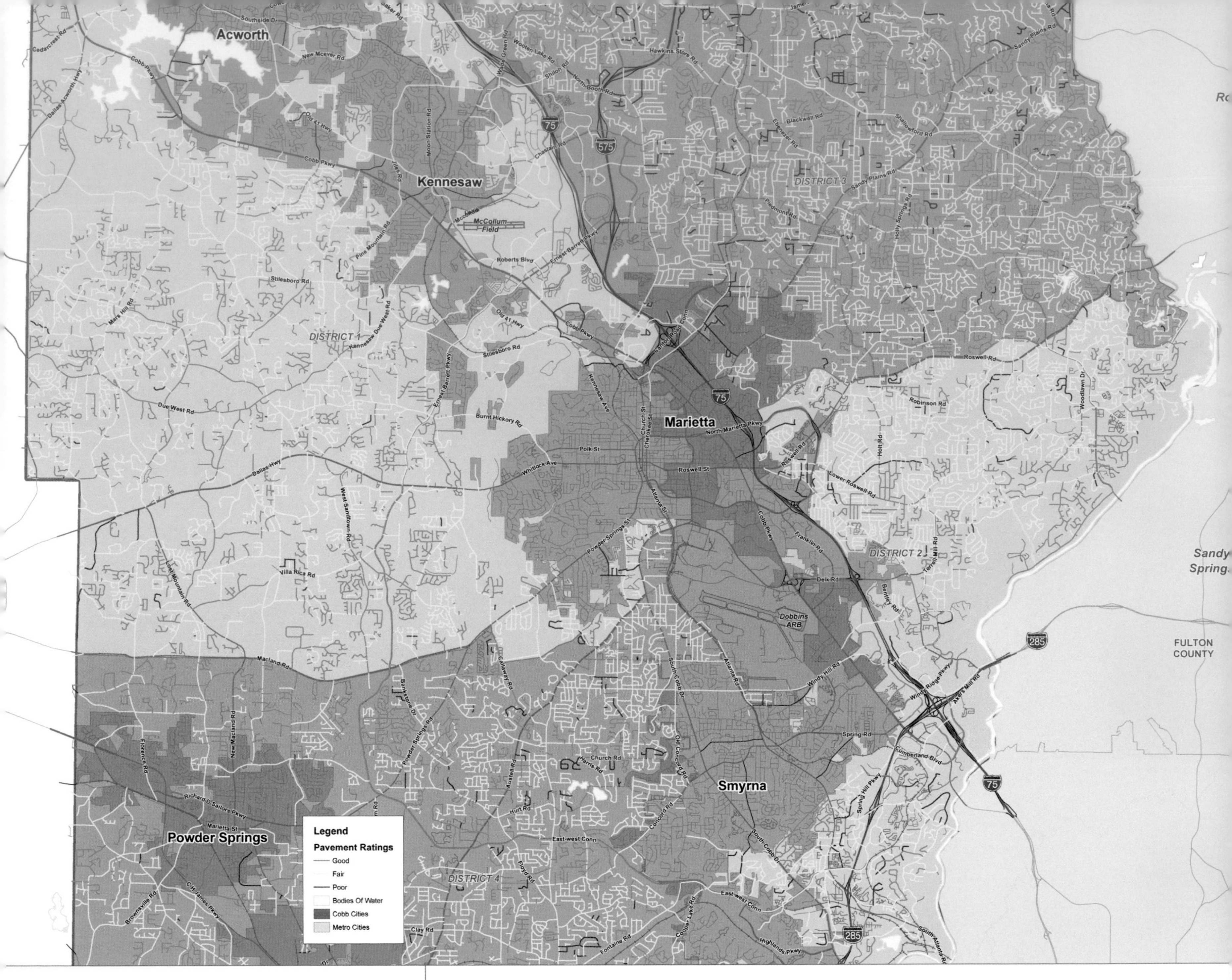

Cobb County Department of Transportation
Marietta, Georgia, USA
By Taylor Busch and Mike Rice

Contact
Lynn Biggs
lbiggs@cobbcounty.org

Software
ArcGIS Desktop, CartêGraph

Printer
HP Designjet 5500ps

Data Source
Cobb County Department of Transportation

Cobb County's Department of Transportation (DOT) integrated its pavement management application with its enterprise transportation layer, CobbETRANS. This map shows the pavement condition ratings for the streets in Cobb's jurisdiction. It helps managers visualize where resurfacing may need to be prioritized.

The DOT develops, manages, and operates the county's transportation systems. Those systems include a vast network of roadways, sidewalks, and pathways; a transit system that provides public transportation services; and a general aviation airport that serves business and recreational flying needs.

Courtesy of Cobb County.

Cobb County Pavement Ratings

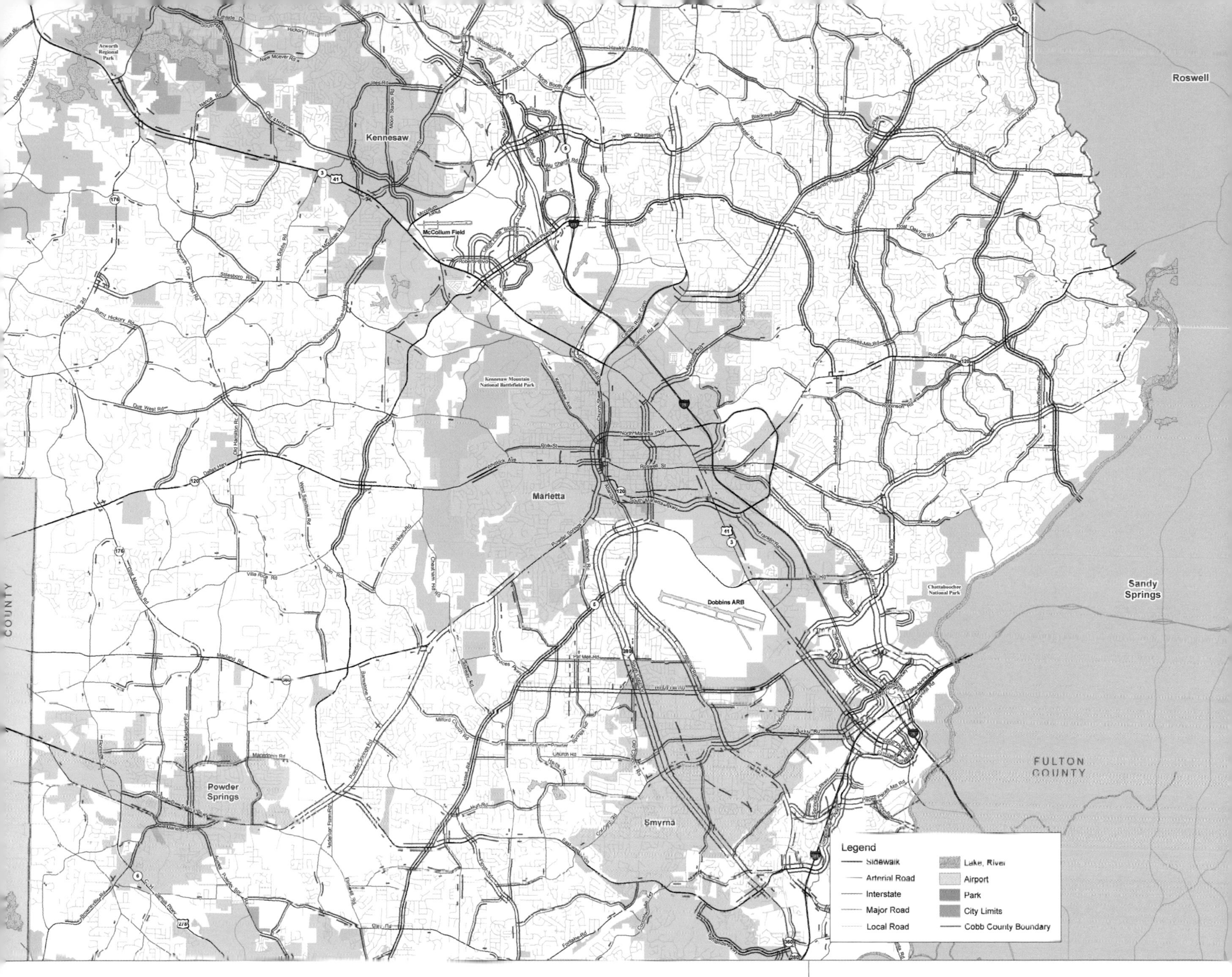

With limited funds and over 500 outstanding requests for new sidewalks, it is important to know where sidewalks exist, where sidewalks are to be built, and where the gaps exist. This map shows the inventory of existing sidewalks on thoroughfare roads in Cobb County and is maintained in the county's enterprise transportation layer CobbETRANS.

The Department of Transportation develops, manages, and operates the county's transportation systems. Those systems include a vast network of roadways, sidewalks, and pathways; a transit system that provides public transportation services; and a general aviation airport that serves business and recreational flying needs.

Courtesy of Cobb County.

Cobb County Department of Transportation
Marietta, Georgia, USA
By Rebecca Snyder

Contact
Lynn Biggs
lbiggs@cobbcounty.org

Software
ArcGIS Desktop, CartêGraph

Printer
HP Designjet 5500ps

Data Source
Cobb County Department of Transportation

The Centre for GIS
Doha, Qatar
By The Centre for GIS, Urban Planning & Development Authority, State of Qatar

Contact
Mohamed Hamouda
mhamouda@gisqatar.org.qa

Software
ArcGIS Desktop 9.2, Adobe Illustrator CS2

Printer
HP Designjet 5500ps

Data Source
Qatar's nationwide GIS database hosted by the Centre for GIS

This map shows the immense GIS data in Qatar shared among forty-four agencies coordinated by The Centre for GIS. Each agency is responsible for updating and maintaining its own data. Public data is shared among all the agencies by replicating the agency data from its server in a batch process to a server accessible by all agencies. All agencies use the data in the public server in their day-to-day activities connected by a high-speed multi protocol label switching network.

Effective coordination is maintained among agencies through a national steering committee and regular meetings involving representatives of all the agencies. Data sharing is standardized with all the agencies using the same ESRI software and compatible data dictionaries. This map shows the agencies connected to The Centre for GIS, Urban Planning & Development Authority. Using ArcToolbox, selected data from each agency was clipped to specific areas based on districts in Qatar. Each agency owning the data is labeled in the map along with the data type.

Qatar's Unique Nationwide GIS Data Sharing

Incorporated in 1881 as a thriving silver mining town, the City of Aspen is now recognized for its world-class skiing, shopping, and high-priced real estate. With robust real estate sales, land development, land-use planning, and general public inquiry, it's critical for Aspen to have an easy-to-read zoning map.

Every public and private parcel is mapped within one of the twenty-four adopted zone district classifications, which determine the development use, dimensions, and intensity regulations on the property. Eight different zone district overlays are also depicted as certain areas of the city require additional regulations. The underlying aerial image provides reference of existing development and proximity to other locations.

In addition to the printed map, zoning information featuring such attributes as description, ordinance number, comments, and general zone classification, is also available online. The map is widely used by city departments and the private sector for land administration, permitting, planning, and analysis. The information in this map comes from enterprise GIS layers that are updated on a daily basis.

Courtesy of Mary Lynne Lackner and the City of Aspen, Colorado.

City of Aspen and Pitkin County
Aspen, Colorado, USA
By Mary Lynne Lackner

Contact
Mary Lynne Lackner
mary.lackner@ci.aspen.co.us

Software
ArcGIS Desktop 9.2

Printer
HP Designjet 4000

Data Sources
City of Aspen and Pitkin County GIS database

City of Aspen Zone District Map

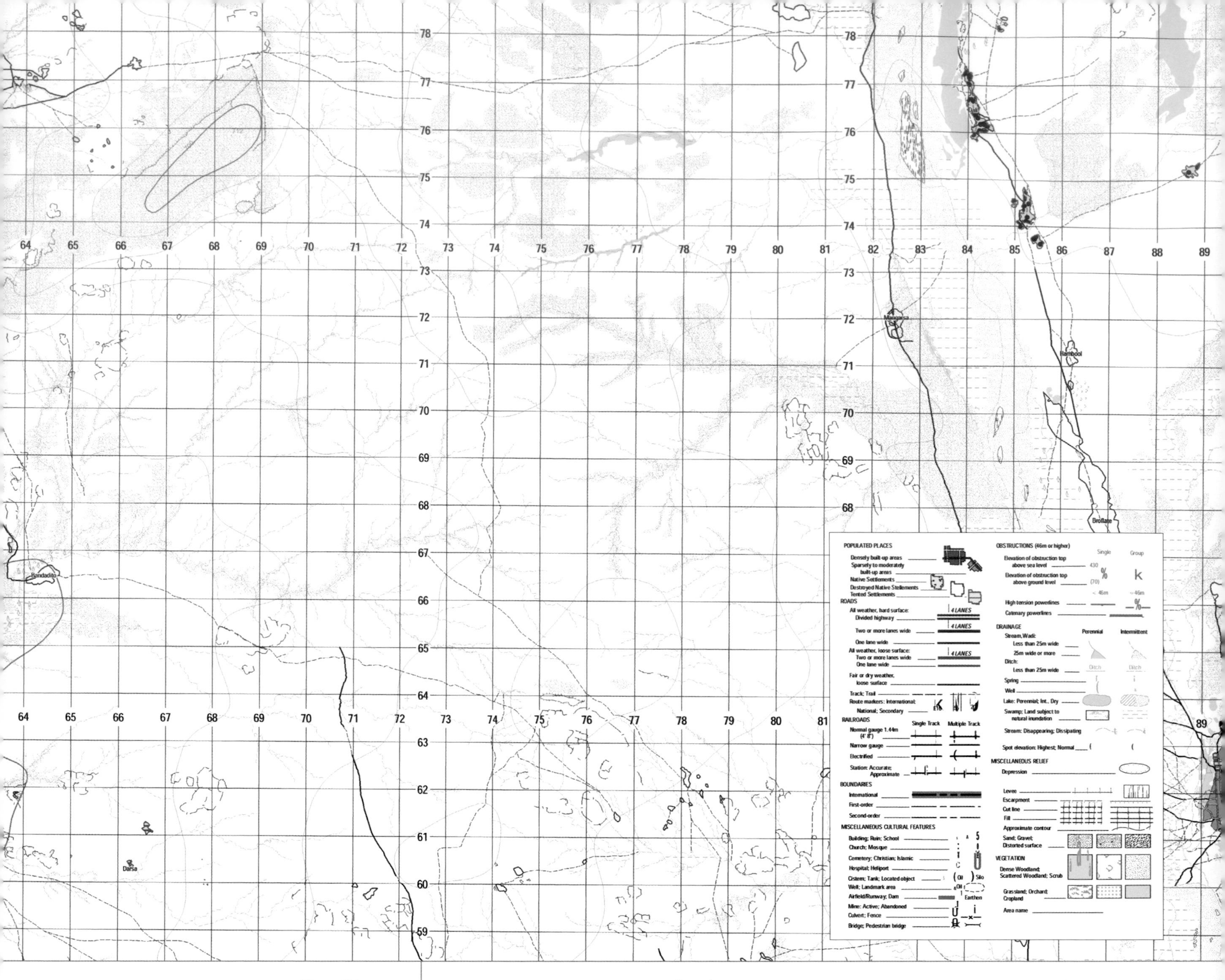

United Nations
New York, New York, USA
By GIS Centre, United Nations Logistics Branch, Brindisi, Italy

Contact
Roy Doyon
doyon@UN.org

Software
ERDAS Imagine, ArcGIS Desktop 9.2

Printer
HP Designjet 5500ps

Data Source
SPOT 5 satellite imagery, 2006

This 1:50,000-scale topographic line map (TLM) of west Darfur, Sudan, is noteworthy because it is new, based on recent SPOT imagery. It portrays basic topographic and cultural features such as roads and settlements, including recently destroyed settlements in a controversial area. The last professionally produced maps of this area were compiled by Soviet Union cartographers in the late 1960s and early 1970s. The TLMs being produced for the Darfur Mapping Project are compliant with the NATO standard Vmap2 geodatabase. An important part of the work, especially feature extraction and map workflow production, was completed using ESRI software, especially PLTS Defense Solution for ArcGIS.

The Darfur Mapping Project is aimed at providing topographic maps at scales ranging from 1:250,000 to 1:25,000 in support of various operational requirements of the United Nations UNAMID peacekeeping mission in Darfur. Accurate and up-to-date mission maps are an ongoing challenge for peacekeeping operations in areas often characterized by rapidly changing human landscapes. Recent trends are to incorporate GIS units within the operational structure of the mission in order to provide mapping support, as well as more focused geospatial awareness. Satellite imagery and map production workflow is the result of a joint collaboration effort between United Nations Cartographic Section (GIS Centre, Brindisi, Italy) and the European Commission, Joint Research Centre, which supported the United Nations in the initial year of work prior to mission start up.

Courtesy of the United Nations, Department of Field Support, Cartographic Section.

North of Baranga, Topographic Quad, Sudan

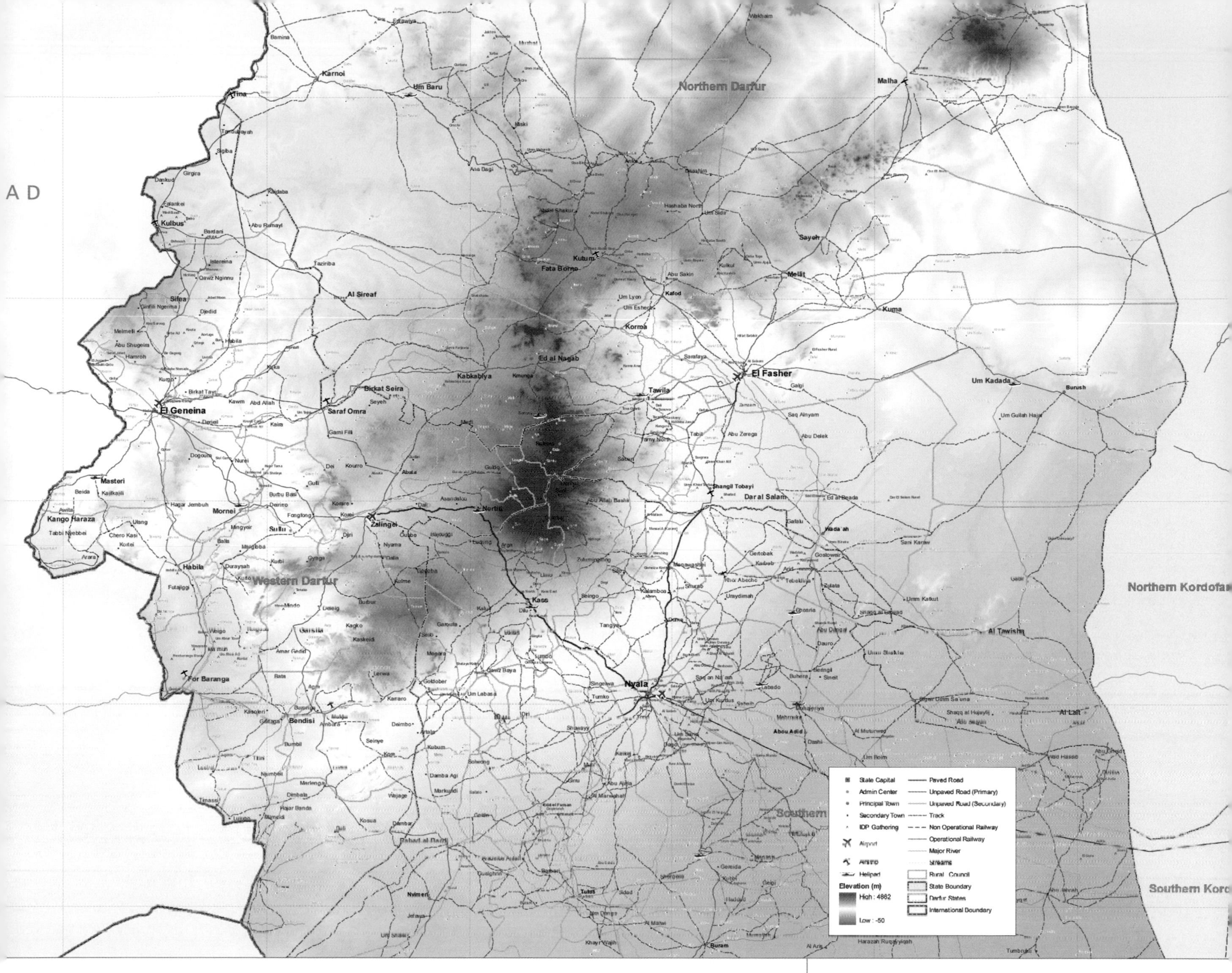

This 1:750,000-scale map of the three Darfur states in western Sudan was designed to help plan assistance efforts for the troubled area. It is a compilation of various sources, including Shuttle Radar Topography Mission Digital Terrain Elevation Data (SRTM DTED), satellite imagery, and locally acquired data for roads and settlements.

It was found to be particularly useful in terms of giving planners an appreciation of the variations of terrain experienced throughout the area, ranging from swampy lowlands to plateaus and higher mountains. Each of these geographic features presents particular problems in terms of access and in turn effects population distribution and density, as evidenced by this product. These factors also influence concentrations of internally displaced persons, a large focus of users of this product.

Once strategies were developed based on this information, larger scale, more detailed maps were developed for individual areas.

Courtesy of the United Nations, Department of Field Support, Cartographic Section.

United Nations
New York, New York, USA
By GIS Section United Nations Assistance Mission in Darfur, El Fasher, North Darfur, Sudan

Contact
Roy Doyon
doyon@UN.org

Software
ERDAS Imagine, ArcGIS Desktop 9.2

Printer
HP Designjet 5500ps

Data Sources
Food and Agriculture Organization, United Nations Joint Logistics Centre, United Nations Mine Action Service, SRTM DTED

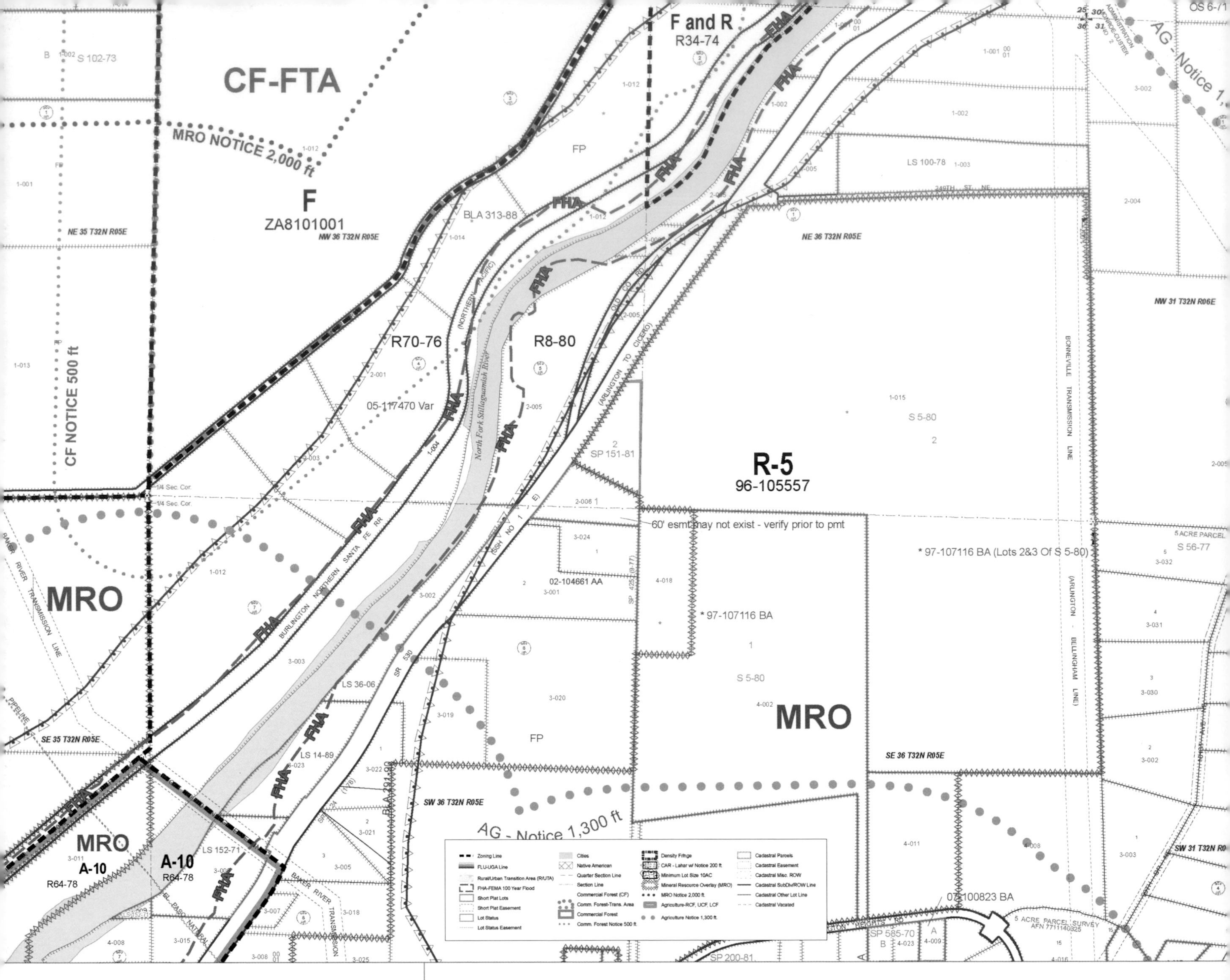

Snohomish County
Everett, Washington, USA
By Cartography/GIS Services

Contact
Carrol Lane
cb.lane@co.snohomish.wa.us

Software
ArcGIS Desktop, ArcGIS Maplex, PLTS MapAtlas

Printer
HP Designjet 1055

Data Sources
Snohomish County Department of Planning and Development Services, Assessor's Office

This map is representative of the Snohomish County Zoning map atlas that includes 1,910 section and quarter-section maps. Over fifty data layers on the maps include zoning, future land use, critical areas, regulations data, subdivision status, and jurisdictional boundaries.

ArcGIS Maplex and PLTS MapAtlas extensions provide map production. ArcMap and ArcServer are used in the management of the feature datasets with registered versioning to allow for multiuser editing.

The atlas is produced by the Snohomish County Department of Planning and Development Services and is used on the permit/zoning counter as a reference for department staff and the public.

Courtesy of Snohomish County, Washington, Department of Planning and Development Services.

Snohomish County Zoning Atlas—Sample Map Section 36 T32N R05E

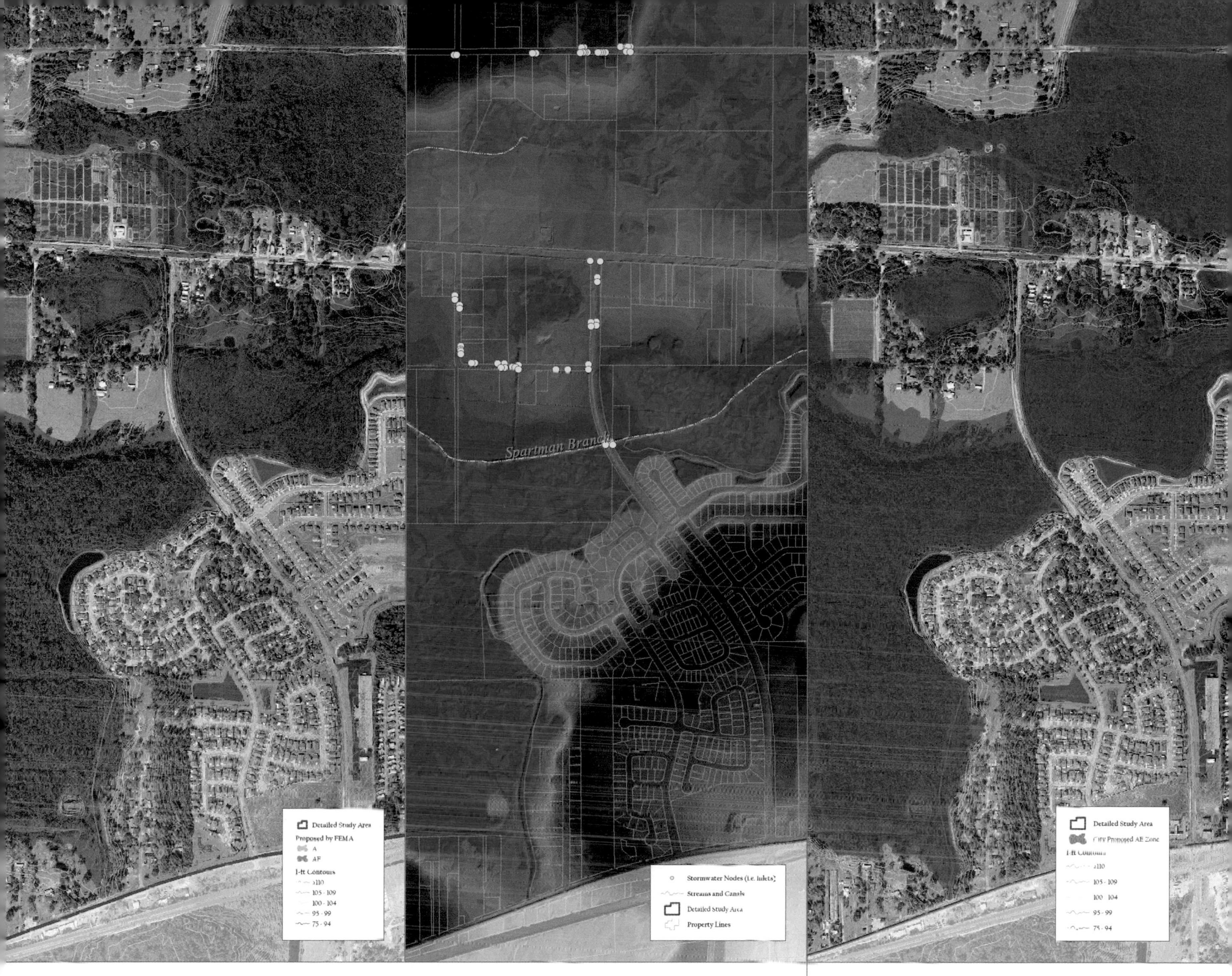

This map was part of an appeal process protesting some of the flood zone determinations in the newly proposed Federal Emergency Management Agency (FEMA) Flood Insurance Rate Maps (FIRM) for 2007. Using drainage modeling, culvert locations, 1-foot contours, lidar data, as-built plans, and lot grading, as well as field measurements and observations, some of the proposed maps for areas throughout the city were appealed. The Sugar Creek/Spartman Branch location was one of these areas. A FEMA review process accepted the City of Plant City's proposed flood zone determination for the area depicted.

Courtesy of Brett Gocka and Zlatko Knezevic, Engineering Division, City of Plant City, Florida.

City of Plant City
Plant City, Florida, USA
By Brett Gocka and Zlatko Knezevic

Contact
Zlatko Knezevic
zknezevic@plantcitygov.com

Software
ArcGIS Desktop, ArcScene

Printer
Canon iPF8000

Data Sources
City of Plant City, Hillsborough County, Southwest Florida Water Management District, Florida Department of Environmental Protection

Adjusting Proposed Flood Insurance Rate Maps (FIRM)

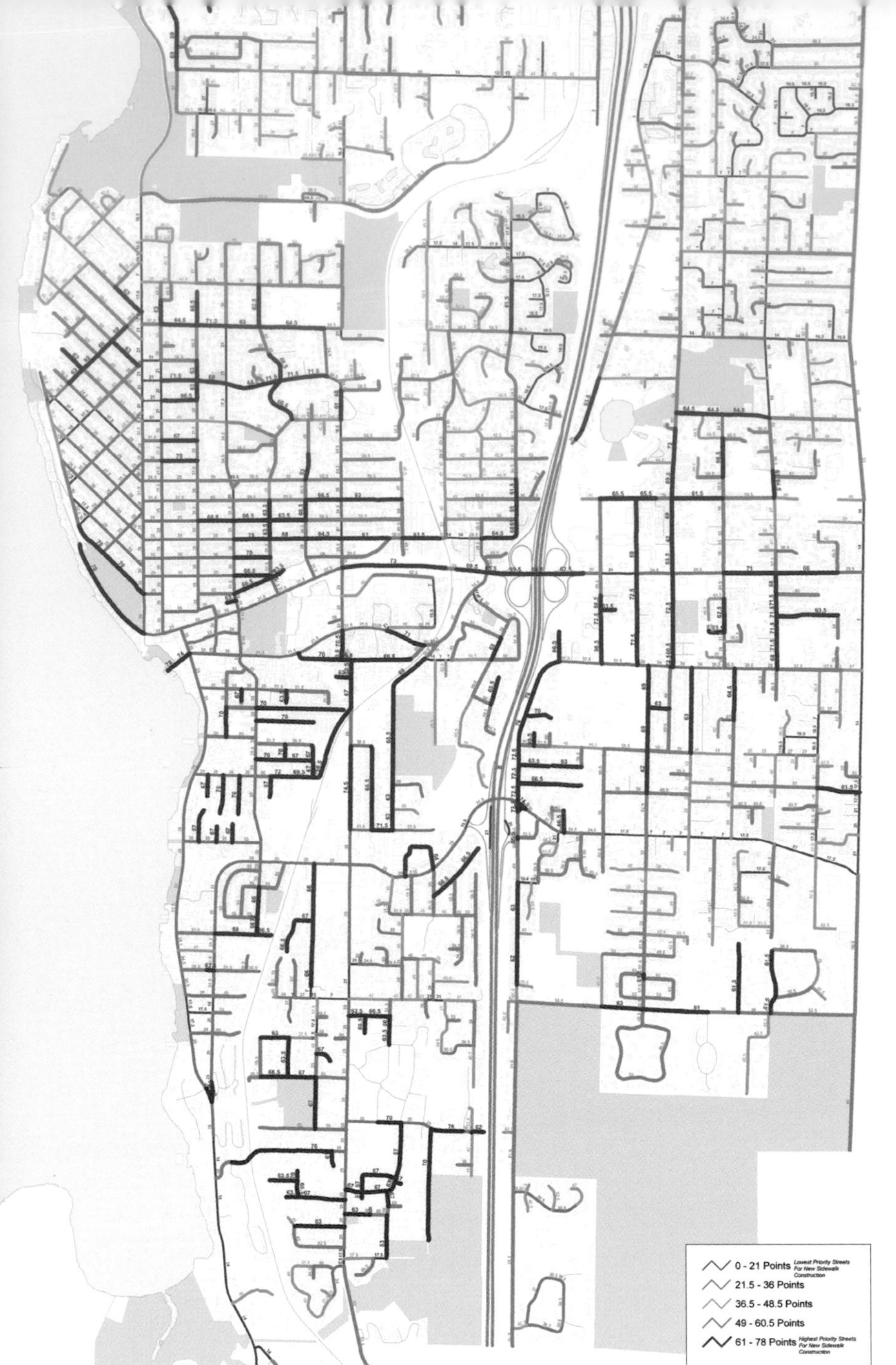

Priority for Sidewalk Construction

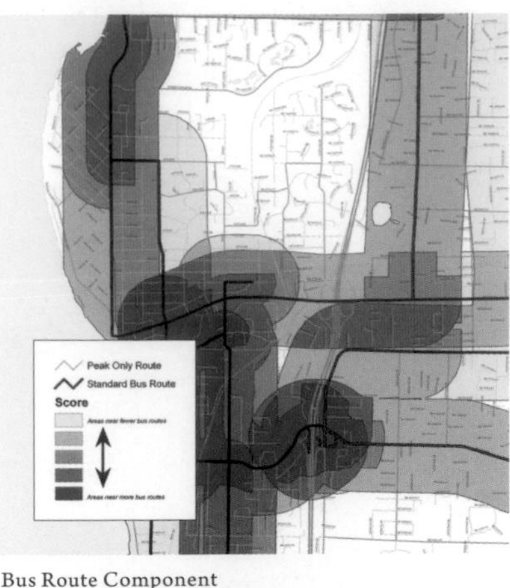

Bus Route Component

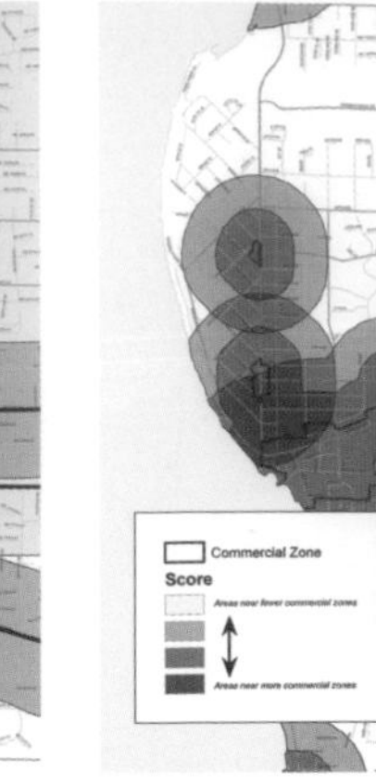

Commercial Component

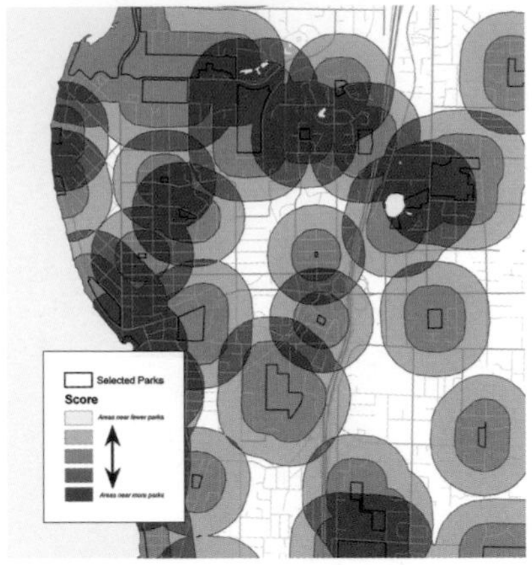

Park Component

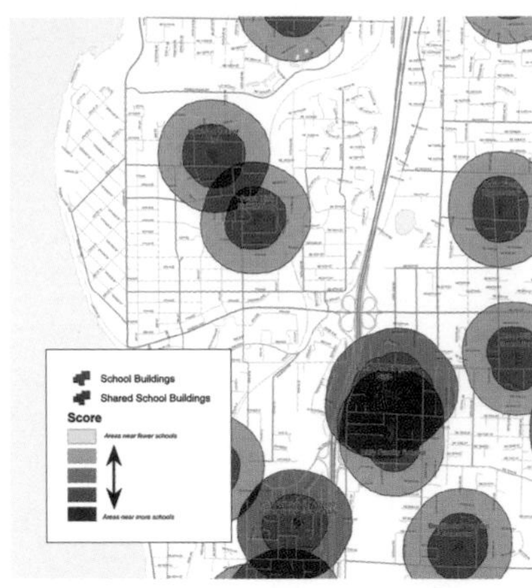

School Component

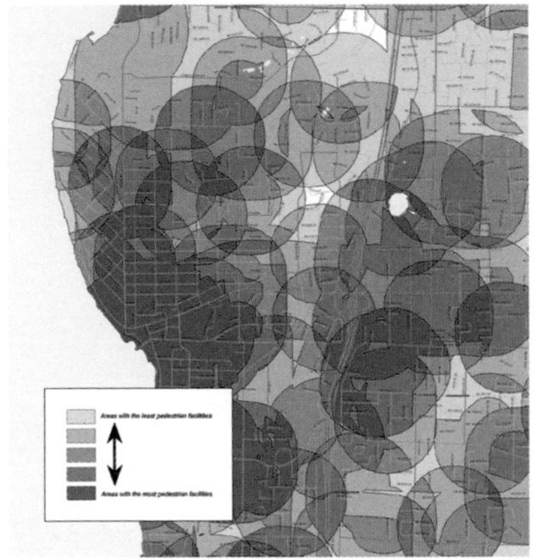

Combined Pedestrian Component Analysis

Sidewalk Completion Status by Street Segment

City of Kirkland
Kirkland, Washington, USA
By David Godfrey and Joe Plattner

Contact
Joe Plattner
jplattner@ci.kirkland.wa.us

Software
ArcGIS Desktop

Printer
HP Designjet 4500

Data Source
City of Kirkland

The City of Kirkland's Public Works Department used ArcGIS software to develop a method for prioritizing new sidewalk construction. Results from a community survey showed that pedestrians are interested in easy access to parks, schools, bus stops, and commercial areas and wanted new sidewalks to fill in gaps and extend the existing sidewalk network.

A three-factor prioritization system was developed to assign a score to each street segment in the city network. Using the ArcGIS Spatial Analyst extension, the proximity to four types of pedestrian-oriented facilities was calculated and then combined into a single measure. Each segment in the city road network was given a score based on the results of this analysis.

In the city's sidewalk inventory geodatabase, each sidewalk segment is identified by the street segment it is adjacent to and by the side of the street it is on. This allowed a quick comparison of the sidewalk inventory against the street segments to identify where gaps in the sidewalk system existed. This information along with the adjacent street segment's functional classification and presence of school walk routes formed the basis for a missing sidewalk score for each street segment.

A score by street segment for existing sidewalk surface type was then combined with the segment level proximity and missing sidewalk scores described above, and that combined measure was mapped. The result was a tool that clearly shows the relative priority for constructing sidewalks on any street in the city.

Active Transportation Plan—Priority Sidewalk Analysis

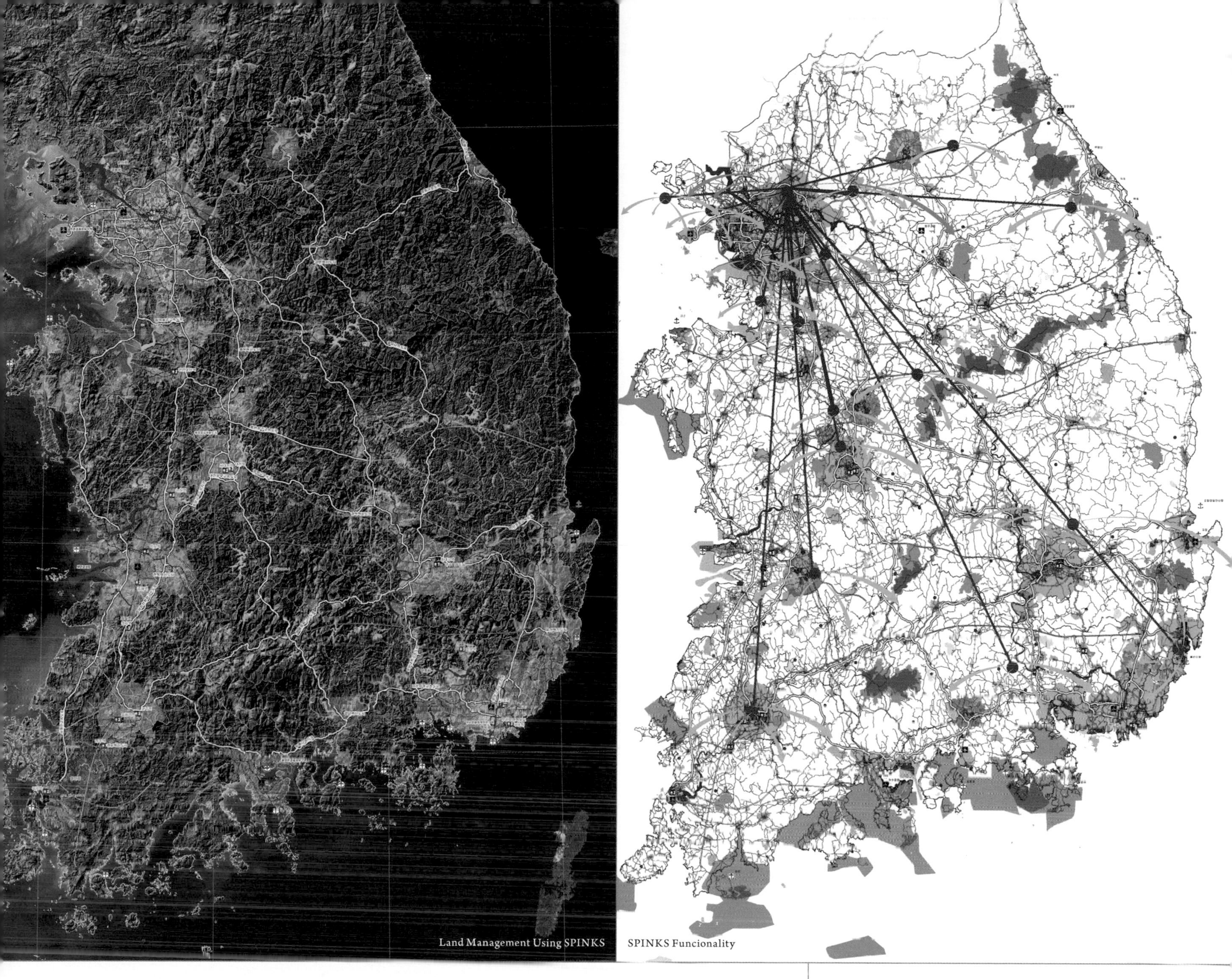

Land Management Using SPINKS

SPINKS Funcionality

SPatial INformation Knowledge System (SPINKS) is a solution to support decision making with spatial analysis at Korea Land Corporation (KLC), whose business domain consists of housing welfare, industrial support, and land management.

Based on spatial data and its attributes, containing terrain, cadastral, and environment information that KLC has built up for a long time, SPINKS has been mainly used for sustainable development and balanced land management. SPINKS enables users to edit and search spatial data on the Web, leveraging ArcGIS geoprocessing tools powered by ArcGIS Server.

The five key roles of SPINKS are developed land management, support for land compensation, support of land banking, land suitability analysis, and policy-making support. With SPINKS, KLC not only gains a geographic advantage, but also makes better decisions with greater efficiency.

Courtesy of Korea Land Corporation.

Korea Land Corporation
Sungnam City, Gyunggi Do, South Korea
By Yong-bum Lee

Contact
Yong-bum Lee
yblee@lplus.or.kr

Software
ArcGIS Server 9.2 Advanced Enterprise

Data Source
Spatial Data Warehouse of Korea Land Corporation

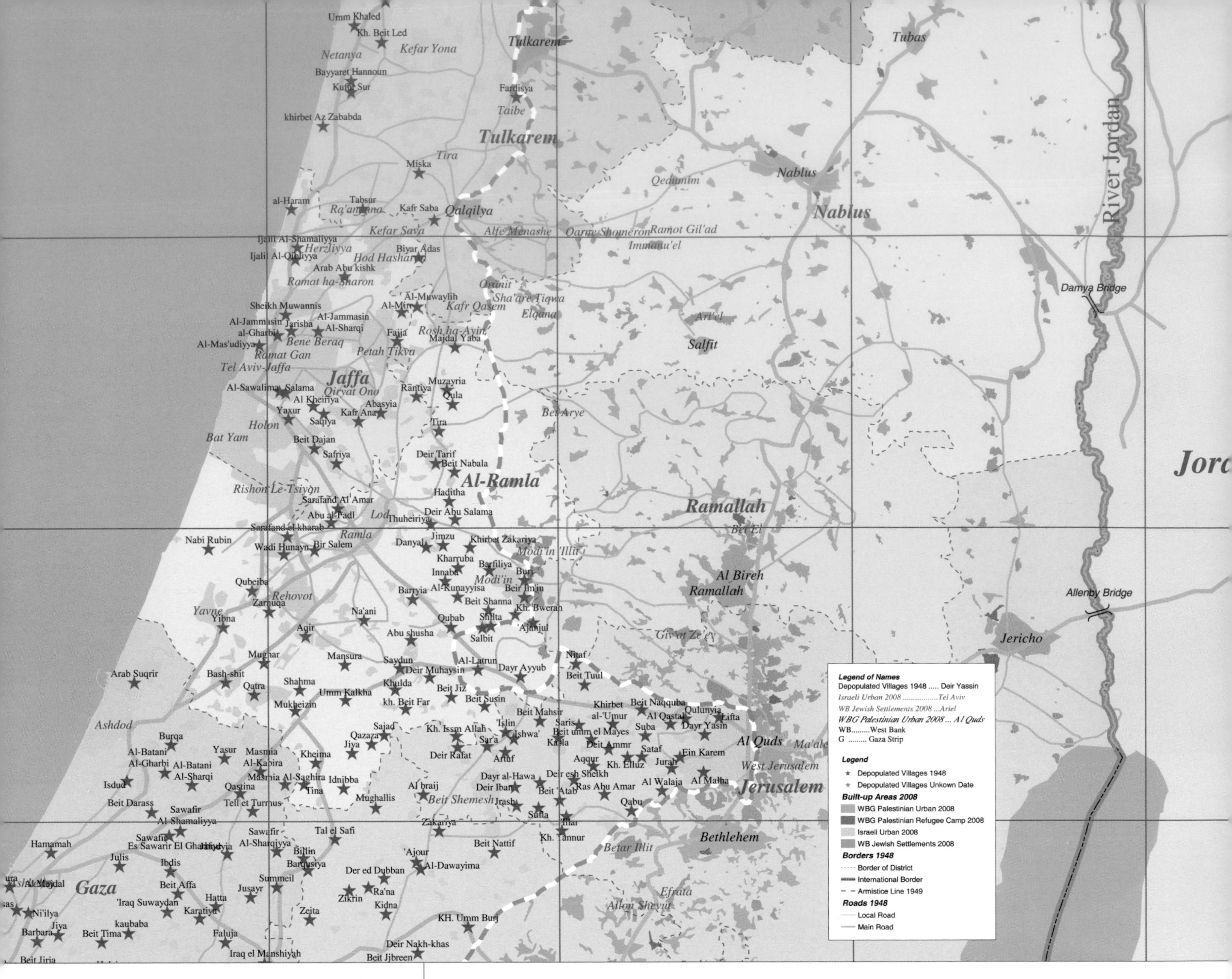

Good Shepherd Engineering & Computing (GSE)/ PalMap, Palestine Mapping Center

Bethlehem, Palestinian National Authority, Palestine

By Michael Younan, Palestine Mapping Center

Contact

Michael Younan

gse@gsecc.com

Software

ArcGIS Desktop 9.2

Printer

Press

Data Sources

PalMap; All That Remains, Walid Khalidi, 1992; The Return Journey, Salman Abu Sitta, Palestine Land Society, 2007; Register of Al Nakbeh 1948, Salman Abu Sitta, the Palestinian Center Al-Awda

This sixtieth anniversary commemorative map shows Palestinian villages and towns depopulated in 1948 compared with current Israeli urban regions. The Palestinian Al-Nakbeh (day of catastrophe) observes the May 15, 1948, forcible evacuation of Palestinians from their homes, villages, and towns upon the creation of Israel. More than six hundred villages and towns were depopulated and destroyed, dispersing Palestinians as refugees worldwide.

This document is part of a new series of "Palestine Alive, Let's Remember," created by Palestine Mapping Center. It is based on historical and recent geographical information and the published works of Palestinian historian Walid Khalidi and Palestinian researcher Salman Abu Sitta. The map and accompanying documents were created with the hope of adding to the knowledge of the Palestinians and their history, and to revive their cultural heritage worldwide.

Courtesy of PalMap, Palestine Mapping Center, a member of GSE.

Map of Palestine, 1948, Depopulated Villages and Towns Compared with Map 2008

This map reflects the areas deemed regionally important by incorporating natural resources available in a GIS database format and the collective expertise of a select group of ecological and park professionals from various federal, state, local, and private organizations.

GIS coverages include soils, slopes, rivers, and streams, wetlands, floodplains, parks, and natural areas, the 2006 Bond Natural Area for Clean Air and Water target areas, Metro's regionally signficant habitat inventory, greenways, and natural hazard data. Additional data from Clark, Yamhill, Marion, and Columbia counties as well as from the Oregon Natural Heritage Program's habitat priorities and The Nature Conservancy's Willamette Valley Ecoregion protection priorities was also used for this project.

These disparate data sources were presented to a panel of participants who were selected for their intimate knowledge of the regional landscape, their grounding in ecological and landscape ecology principles, and their familiarity with Metro's regional growth management and greens-paces program.

The cartographic product is a map of significant natural systems and land patterns that define the quality and character of the region and a diagrammatic concept for the "system" that captures the region's sense of place, allows for resource protection at a larger landscape and ecosystem scale, and helps define where future growth should and should not occur.

Courtesy of Matthew Hampton.

Oregon Metro
Portland, Oregon, USA
By Matthew Hampton and Max Woodbury

Contact
Matthew Hampton
matthew.hampton@oregonmetro.gov

Software
ArcGIS Desktop, Adobe CS2

Printer
HP Designjet 4500ps

Data Sources
Metro Regional Land Information System, ESRI, National Park Service Natural Earth, Nature Conservancy, Oregon Natural Heritage Program

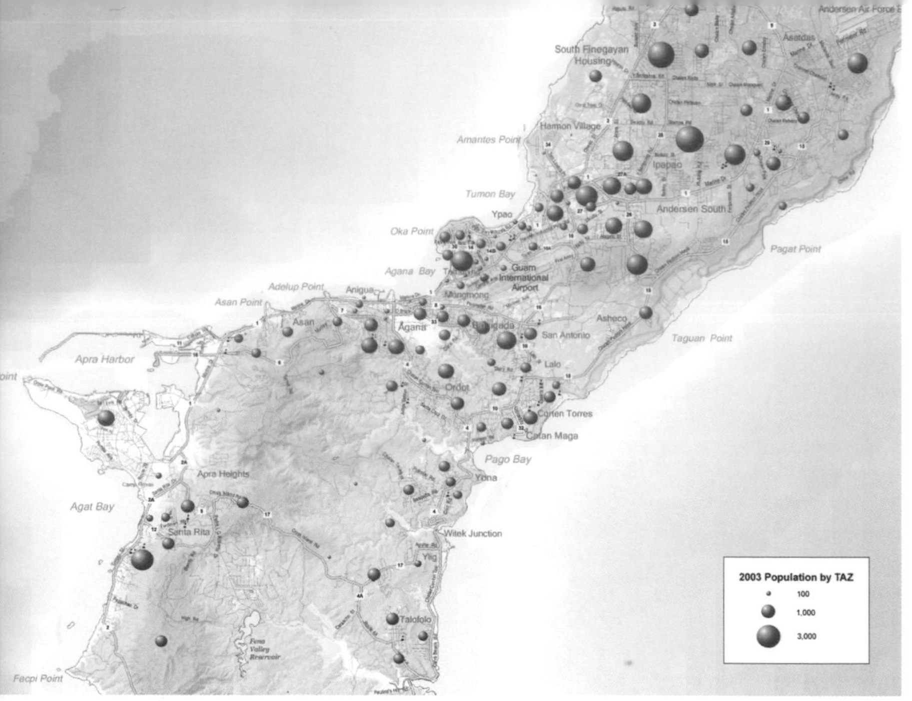

2003 Population by Traffic Analysis Zones

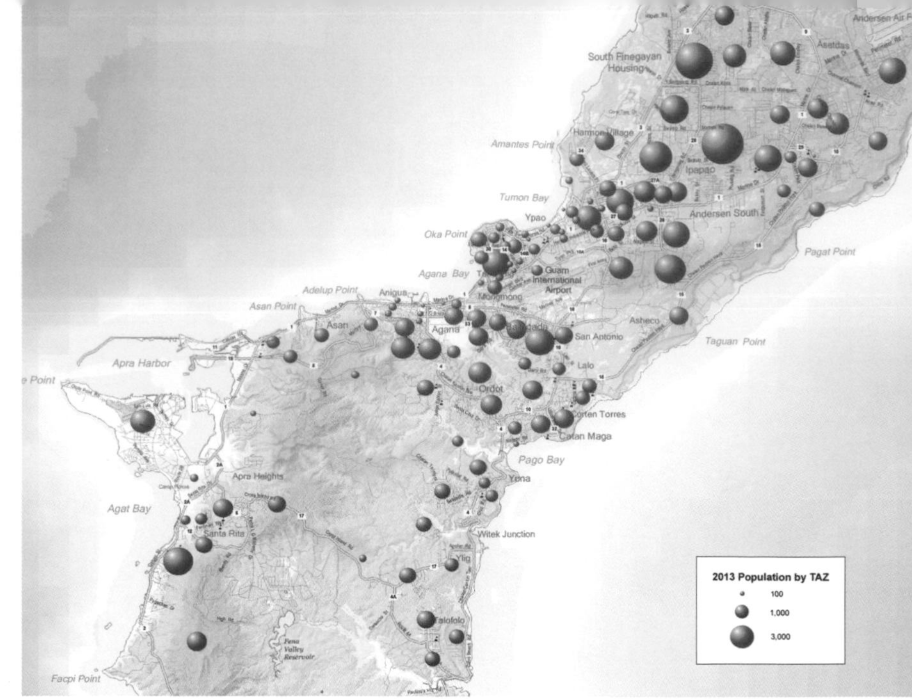

2013 Population by Traffic Analysis Zones

2030 Housing with Naval Base Expansion

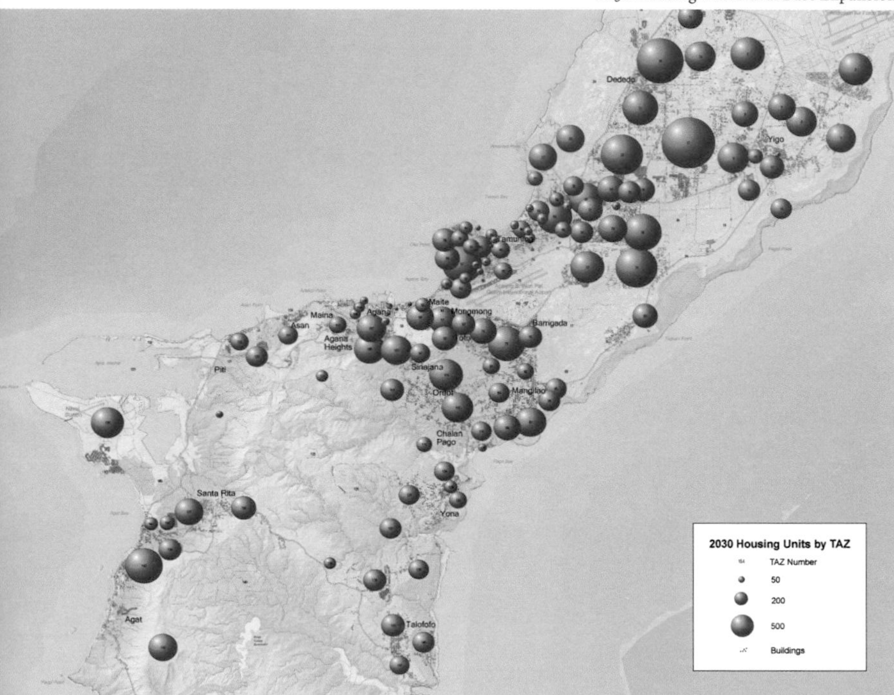

2030 Housing without Naval Base Expansion

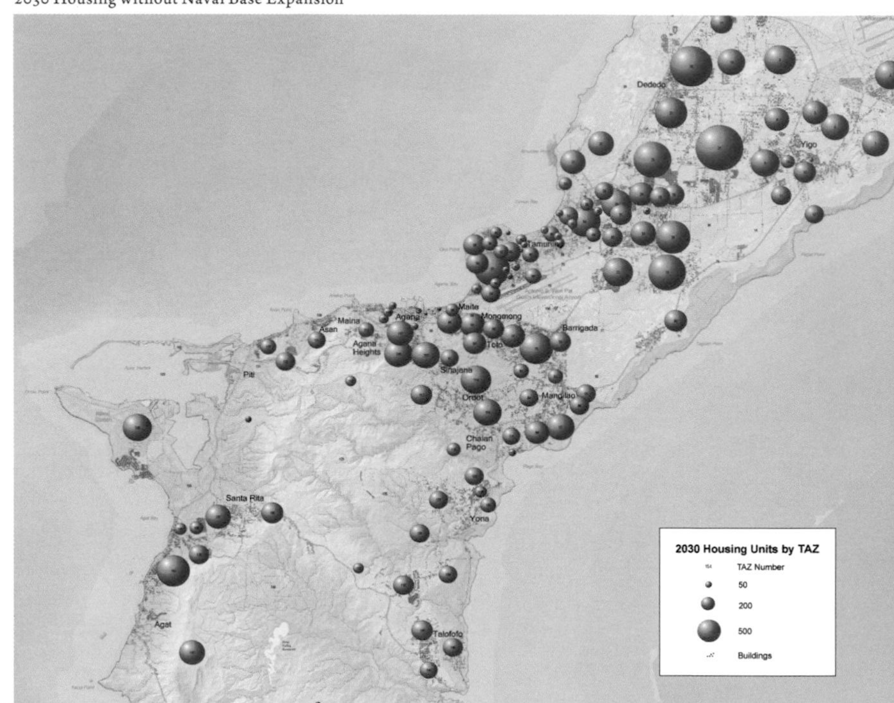

Parsons Corp.
San Jose, California, USA
By Eric Coumou

Contact
Eric Coumou
eric.coumou@parsons.com

Software
ArcGIS Desktop 9.3

Printer
HP Designjet 1055cm

Data Sources
U.S. Census Bureau, Guam GIS Department

Using U.S. Census data, the distribution of dwelling units is mapped for the island of Guam. This map will help transportation planners manage the growth of Guam upon the relocation of several U.S. military bases.

Over the next several years, the United States will relocate thousands of military personnel from Okinawa, Japan, to the island of Guam. The relocation of thousands of marines and their families will have a major impact on the island's infrastructure. Housing, schools, and services will have to be built. New roads and new utilities will be needed, and existing roads will have to be upgraded.

These maps were part of a series of maps created to show the distribution of different demographic variables depending on various development scenarios on the island of Guam. Using traffic modeling software and inputs from Traffic Analysis Zone data, employment and housing is forecast for future years. ArcGIS is used to manage and symbolize these distributions, and gives planners a good way to visualize the development alternatives.

Courtesy of Parsons Corp.

Island of Guam Demographics

This map was compiled to illustrate the risks from overbank and inundation flooding within the United Nations peacekeeping mission area of Timor-Leste. As part of the volcanic chain of islands comprising the Indonesian archipelago, Timor-Leste is characterized by steep-sided mountains, a monsoon climate, and subsistence agriculture. Unsound agricultural practices increase soil erosion, which in turn heightens the probability of flooding, especially during the rainy season. It is not uncommon in this environment to discover mountain side roads partially collapsed or completely washed away.

This map provides mission staff with a variety of information. Basic trip planning can be done on a day-to-day basis. For logistical assistance, it indicates areas where heavy vehicle traffic should be avoided or where road construction projects can be implemented. It also shows helicopter landing sites for both routine and emergency tasks.

GIS units in peacekeeping missions are often called upon to provide mapping support for a variety of projects. Traffic planning, military deployment, population censuses, election planning and results, security and evacuation routes and planning, and developmental assistance project planning are some of the mapping topics that often arise on such missions.

Courtesy of the United Nations, Department of Field Support, Cartographic Section.

United Nations
New York, New York, USA
By GIS Section, United Nations Mission in Timor-Leste

Contact
Roy Doyon
doyon@UN.org

Software
ArcGIS Desktop 9.2

Printer
HP Designjet 5500ps

Data Source
Historical 2001 data from the National Disaster Management Office

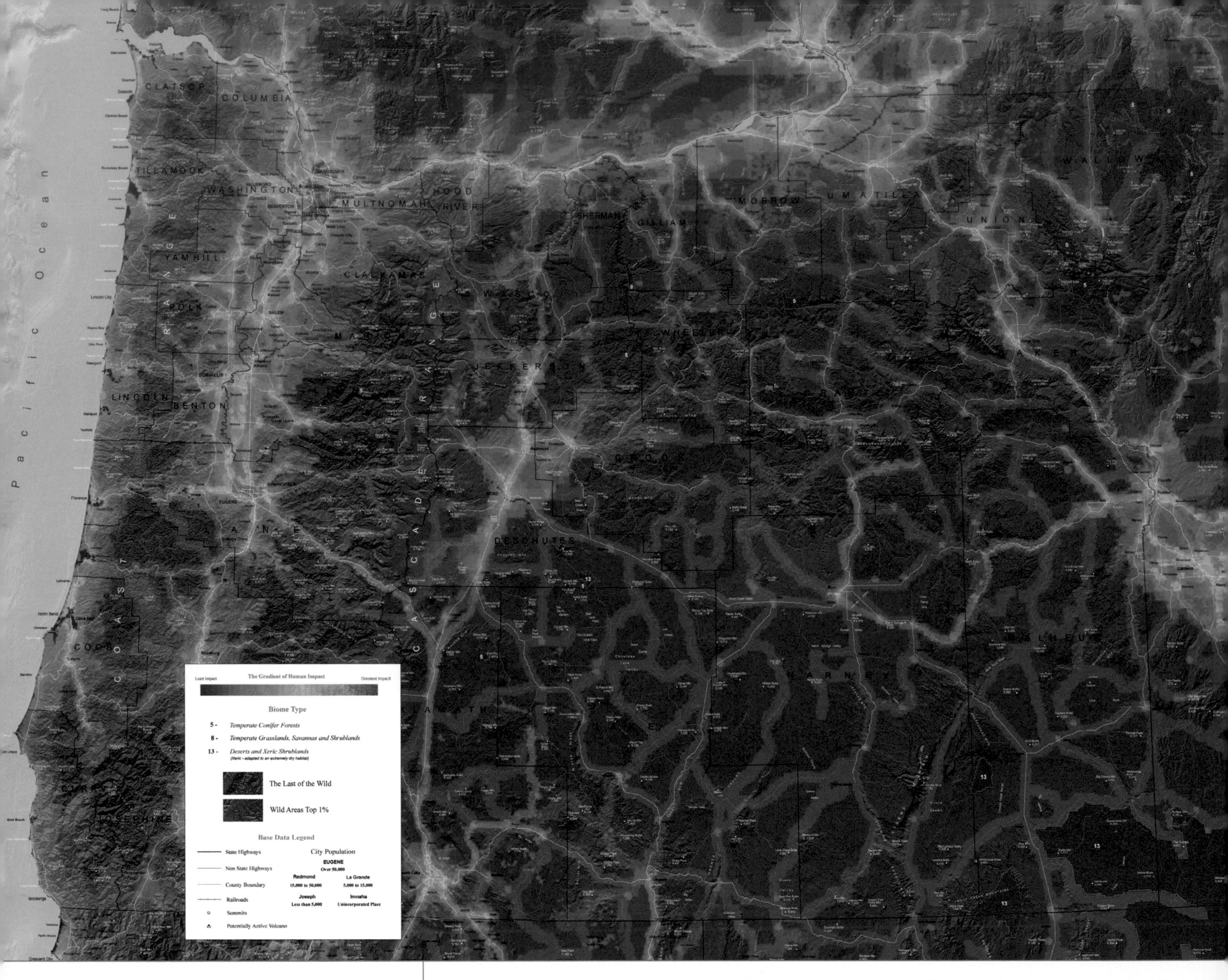

Oregon Department of Transportation
Salem, Oregon, USA
By Ryan R. Johnson

Contact
Ryan R. Johnson
Ryan.R.Johnson@odot.state.or.us

Software
ArcGIS Desktop 9.2, Adobe Photoshop CS2

Printer
HP Designjet 5500ps

Data Sources
Wildlife Conservation Society, Center for International Earth Science Information Network at Columbia University, Oregon Department of Transportation, National Oceanic and Atmospheric Administration, U.S. Geological Survey

Transportation is one of the most significant factors of human influence on our planet. This map of the human footprint shows the gradient of human impact within the state of Oregon. This map is a unique view of human impact that demonstrates the Oregon Department of Transportation's (ODOT) dedication to renewable energy.

It takes 45,000 megawatt hours of electricity annually to run Oregon's state transportation system. This energy is used for signals, illumination, buildings, ramp metering, and more. Historically this energy comes from mostly nonrenewable sources. ODOT supports efforts to reduce greenhouse gas emissions and is planning for the transition to alternative, renewable fuels that will be required for the future.

Oregon's governor has directed state agencies to secure 100 percent of their electricity from renewable sources, and ODOT is responding by developing the nation's first solar highway. With 16,000 lane miles of right of way and many other properties under its ownership, ODOT buildings and lands provide a ready asset for the development of solar energy. ODOT also has active projects involving electric vehicle charging stations, alternative fueling sites, and an environmental data management system to help preserve natural and cultural resources.

Courtesy of the Oregon Department of Transportation, Geographic Information Services.

Reducing the Impact of Transportation on the Human Footprint

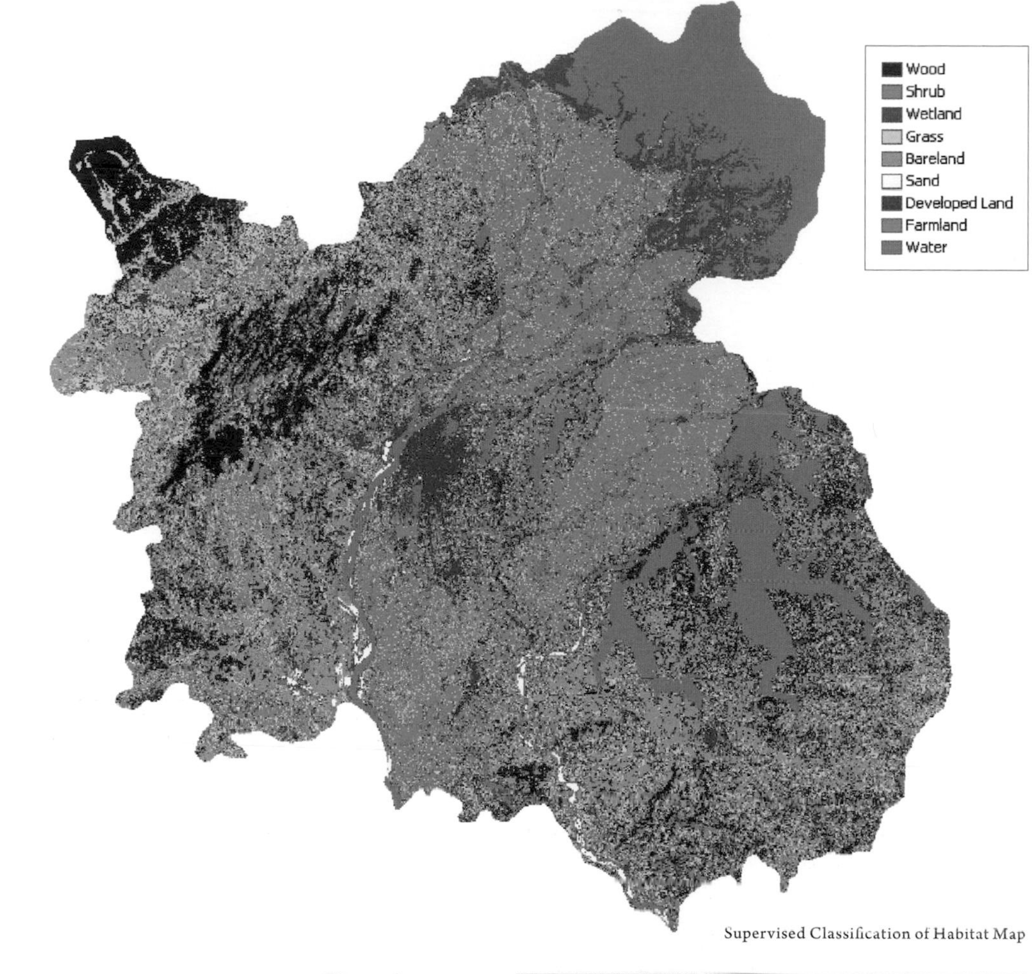

Supervised Classification of Habitat Map

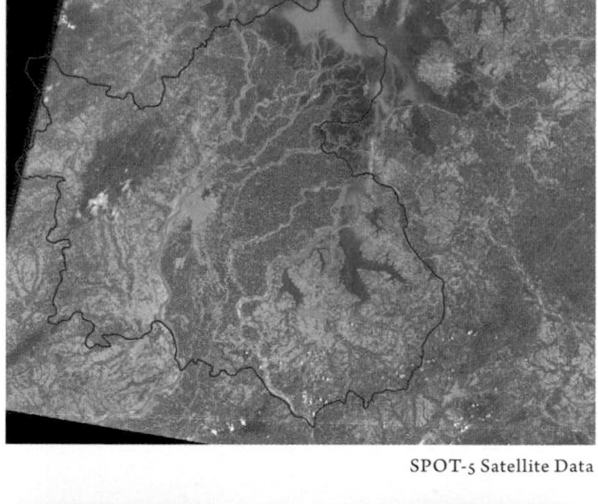

SPOT-5 Satellite Data

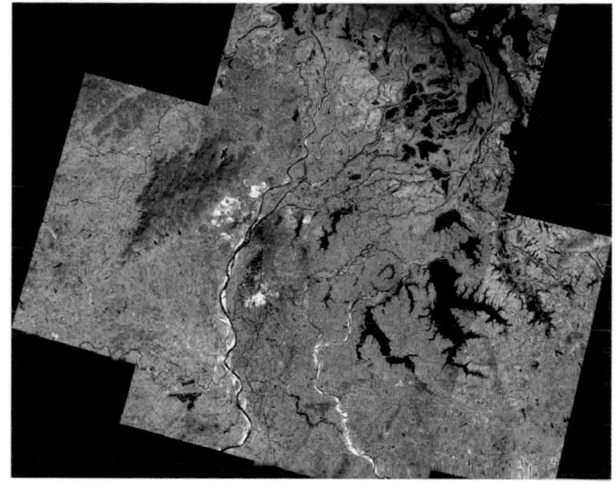

Landsat 7 Satellite Data

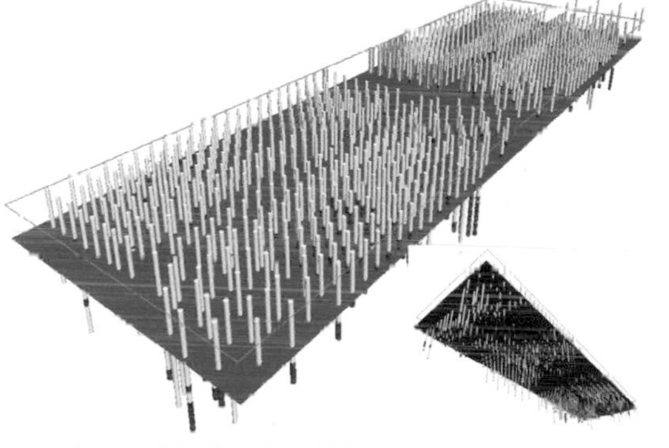

Underground Geological Model for Logistic Center Development

Landslide Suceptibility Modeling

Visualization of Prospective Urban Development

This map summarizes some of the major GIS developments in the past few years of AECOM Asia. It illustrates the customization work done by using the MapObjects and ArcObjects techniques. Customized GIS applications for habitat mapping, natural landslide searching, air pollutant monitoring, and utility map printing were developed to facilitate engineering work. Also presented are 3D geological and landscape simulation results for different study areas in mainland China and Hong Kong.

Courtesy of AECOM Asia.

AECOM Asia
Hong Kong, Hong Kong, China
By Eric Yau

Contact
Eric Yau
eric.yau@aecom.com

Software
ArcGIS Desktop, ArcGIS Spatial Analyst, ArcGIS 3D Analyst

Printer
HP Deskjet 1050c

Data Sources
SPOT 5, Landsat 5, Hong Kong Lands Department, Hong Kong Civil Engineering and Development Department

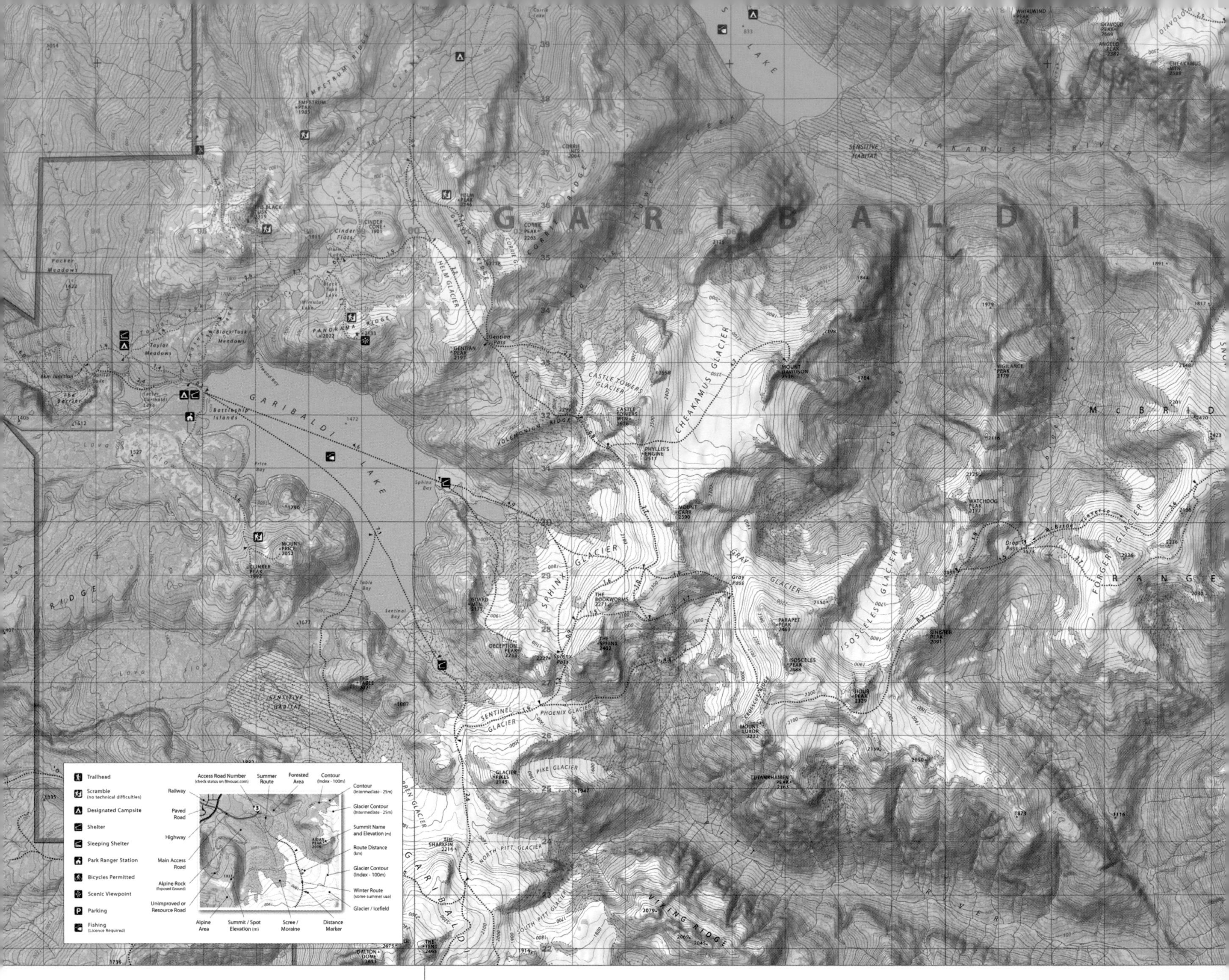

Clark Geomatics Corporation
North Vancouver, British Columbia, Canada
By Jeff Clark

Contact
Jeff Clark
jeff@clarkgeomatics.ca

Software
ArcGIS Desktop, Adobe Photoshop CS2, Adobe Illustrator CS2

Printer
Epson 7800

Data Sources
Landsat 7 ETM+ (geogratis.ca), Canadian digital elevation data (geobase.ca), road centreline network (GIS Innovations Inc.), Canadian National Topographic Database

The map of Garibaldi Park in British Columbia, Canada, was designed and produced using a combination of state-of-the-art satellite image processing methods, geographic information analysis, and classic high-mountain cartographic techniques.

It features full-color, shaded relief, 1:50,000-scale topographic map with trails and backcountry routes, 20-meter contours, terrain types (rock, forest, alpine, ice, etc.), clearly marked access roads and trailheads along with a cross-referenced information guide for both hikers and skiers. The cover of the map was also produced at 1:50,000 scale to allow people traversing the Garibaldi Névé to use the map in its folded state.

Bivouac Backcountry Series: SW Garibaldi Park, BC

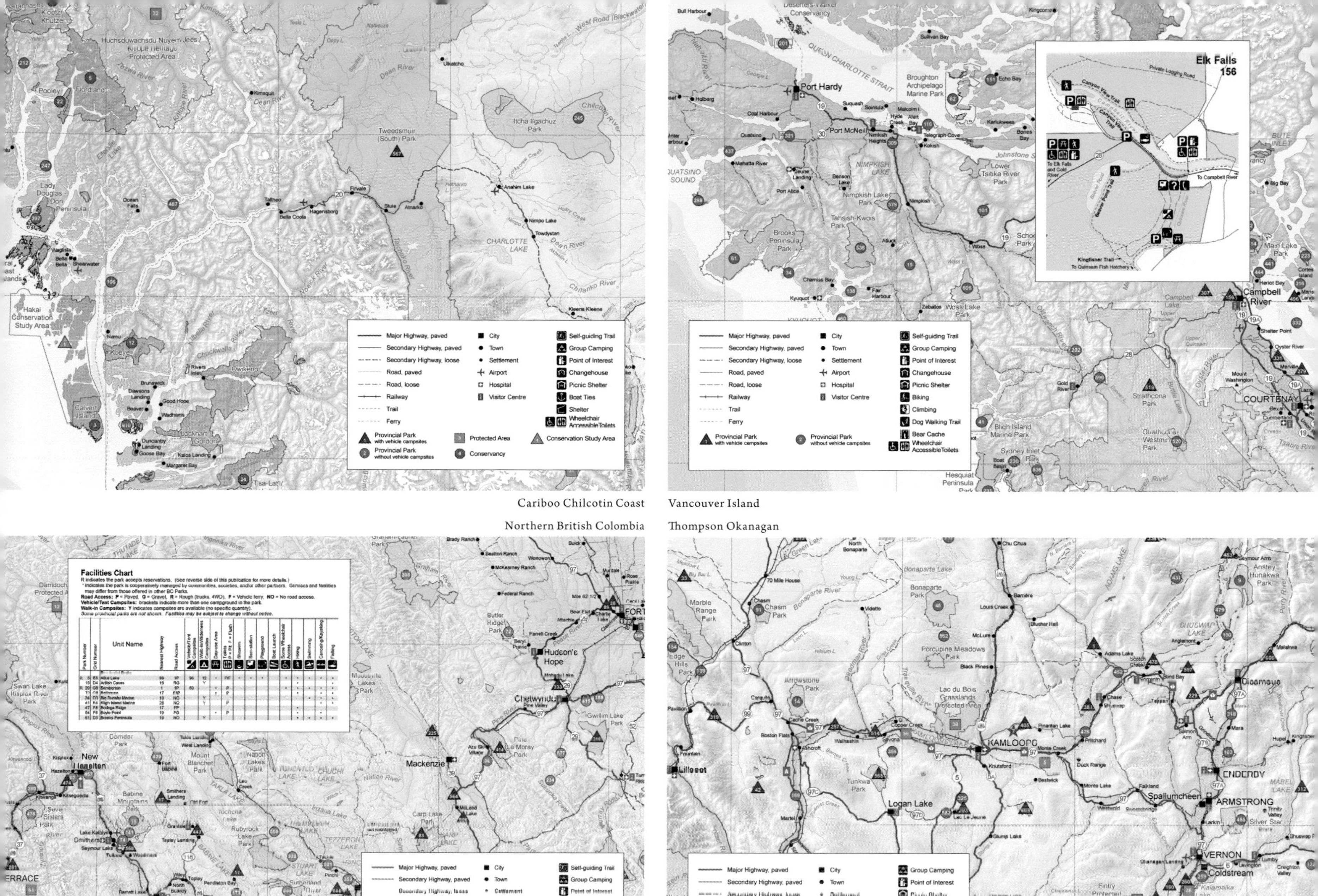

Cariboo Chilcotin Coast

Vancouver Island

Northern British Colombia

Thompson Okanagan

The provincial parks maps project included Thompson Okanagan, Kootenay Rockies, Cariboo Chilcotin Coast, Vancouver Coast and Mountains, Vancouver Island, and Northern British Columbia. The project consisted of designing and creating the map sides of publication-quality, regional visitors' guides of British Columbia. The maps show the locations of provincial protected areas, highlight individual parks by the use of inset maps, and list the facilities available in the parks.

The format of the previous set of maps, which were developed in the early 1990s, was used as a general guide, and the data frame from an ArcGIS Desktop project of a recently created, large-scale, provincial wall map was used as a data layer and symbology template.

A major challenge with the main map included locating, acquiring, creating, and editing an appropriate roads file for the project. The challenge posed was not only to show enough roads but also not to show too many roads.

The final step for each map involved importing the facilities chart spreadsheet below the legend area. This was complicated because only the raw text from the spreadsheet would import properly. The font color, borders, and fill color would not import completely, so those elements had to be created manually in ArcMap.

Courtesy of GeoBC Spatial Anlaysis Branch, Integrated Land Management Bureau, Province of British Columbia.

GeoBC Spatial Analysis Branch
Victoria, British Columbia, Canada
By Tristan Joslin (GeoBC Spatial Analysis Branch) and Kim Reid (Ministry of Environment, BC Parks)

Contact
Carol Ogborne
Carol.Ogborne@gov.bc.ca

Software
ArcGIS Desktop 9.2

Printer
HP Designjet 4500

Data Sources
1:2,000,000 NTS data for river, lakes, and ocean annotation; 1:20,000 parks and protected areas; 1:250,000 NTS glaciers; 1:20,000 Integrated Land Management Bureau Terrain Resource Information Management digital elevation model and basemap layers; various in-house datasets (airports, highway signs, First Nation reserves)

Park Maps of British Columbia

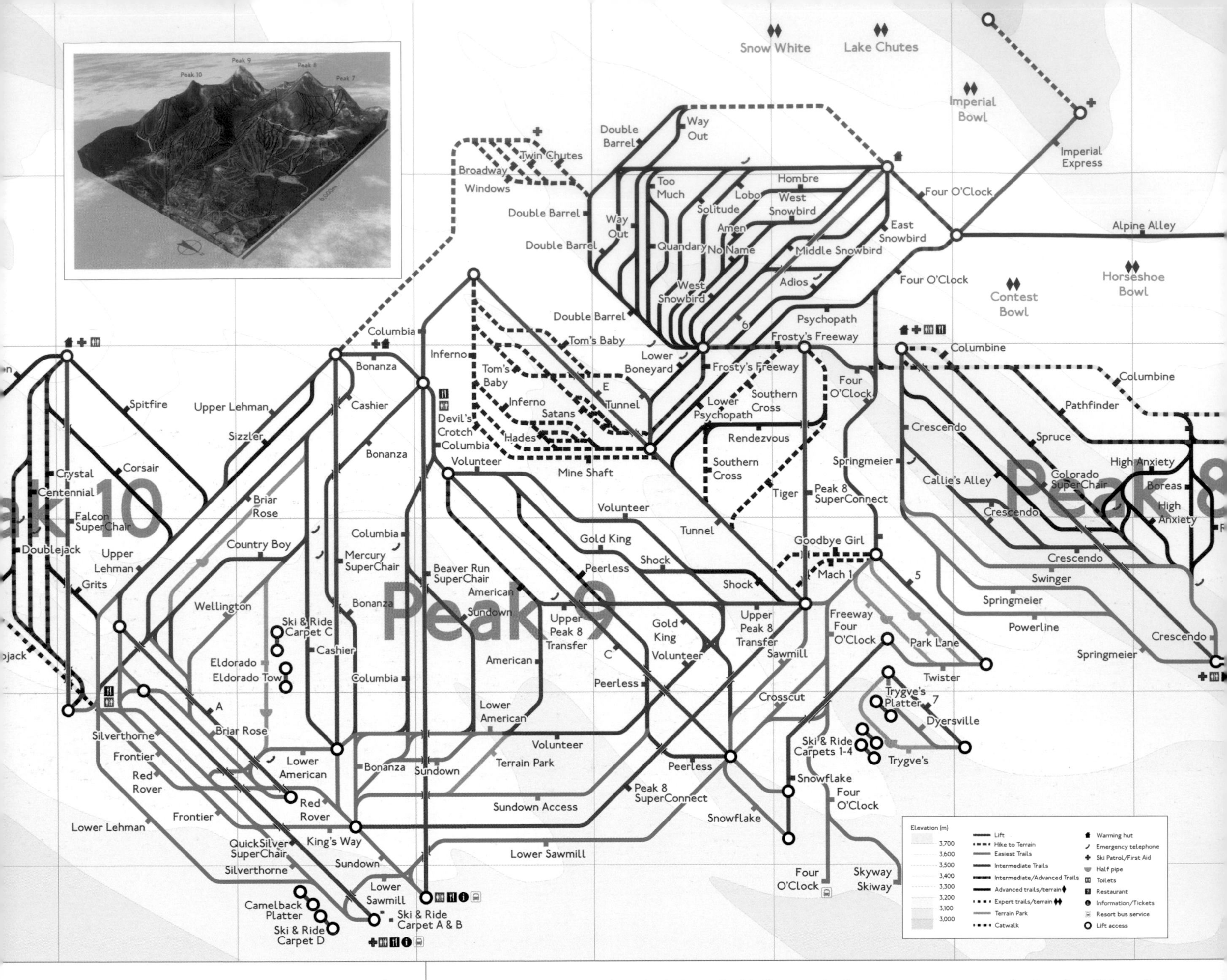

Kingston University London
Kingston Upon Thames, Surrey, United Kingdom
By Kenneth Field

Contact
Kenneth Field
ken.field@kingston.ac.uk

Software
ArcGIS Desktop 9.3, ArcGIS Network Analyst, ArcGIS Schematics, ArcGIS 3D Analyst, ArcGIS Spatial Analyst, ArcScene, Natural Scene Designer

Printer
HP Designjet

Data Sources
GPS tracklogs, U.S. Geological Survey data elevation model

Trail maps for winter sports often take the form of highly illustrative landscape panoramas, overprinted with trails and lifts to allow skiers and snowboarders a mechanism for navigation. However, they often require imaginative interpretation to understand the various on-mountain divides and terrain characteristics crucial to orientation and safety.

This map, which illustrates the 176 trails at Breckenridge Ski Resort in Colorado, takes a radically alternative approach to the traditional trail map by reducing a mountain, its trails, and lifts to topological primitives. The map takes a similar approach to that of the famous London Underground map, and many other transit maps, making the network of trails and lifts visible and immediate, thus enhancing a users' ability to navigate the interconnectivity of lifts and trails combined with the level of difficulty of the trails. Representing the information as a network makes it far easier to determine a route from the top of a lift back down the mountain according to ability.

In a fashion similar to many transit maps, the network is constructed from horizontal, vertical, and 45-degree lines. Congested areas are exaggerated, and sparse areas truncated. Movement between the lift and trail network is indicated by a "station junction" symbol, and trail and lift names are included. Line symbols mirror those routinely used on trail maps to identify the different categories of difficulty (green for easy, black for difficult, etc.). The map won the best software integration and best overall map categories in the Map Gallery competition at the 2008 ESRI European, Middle East, and Africa User Conference.

Breckenridge Ski Resort Topological Trail Map

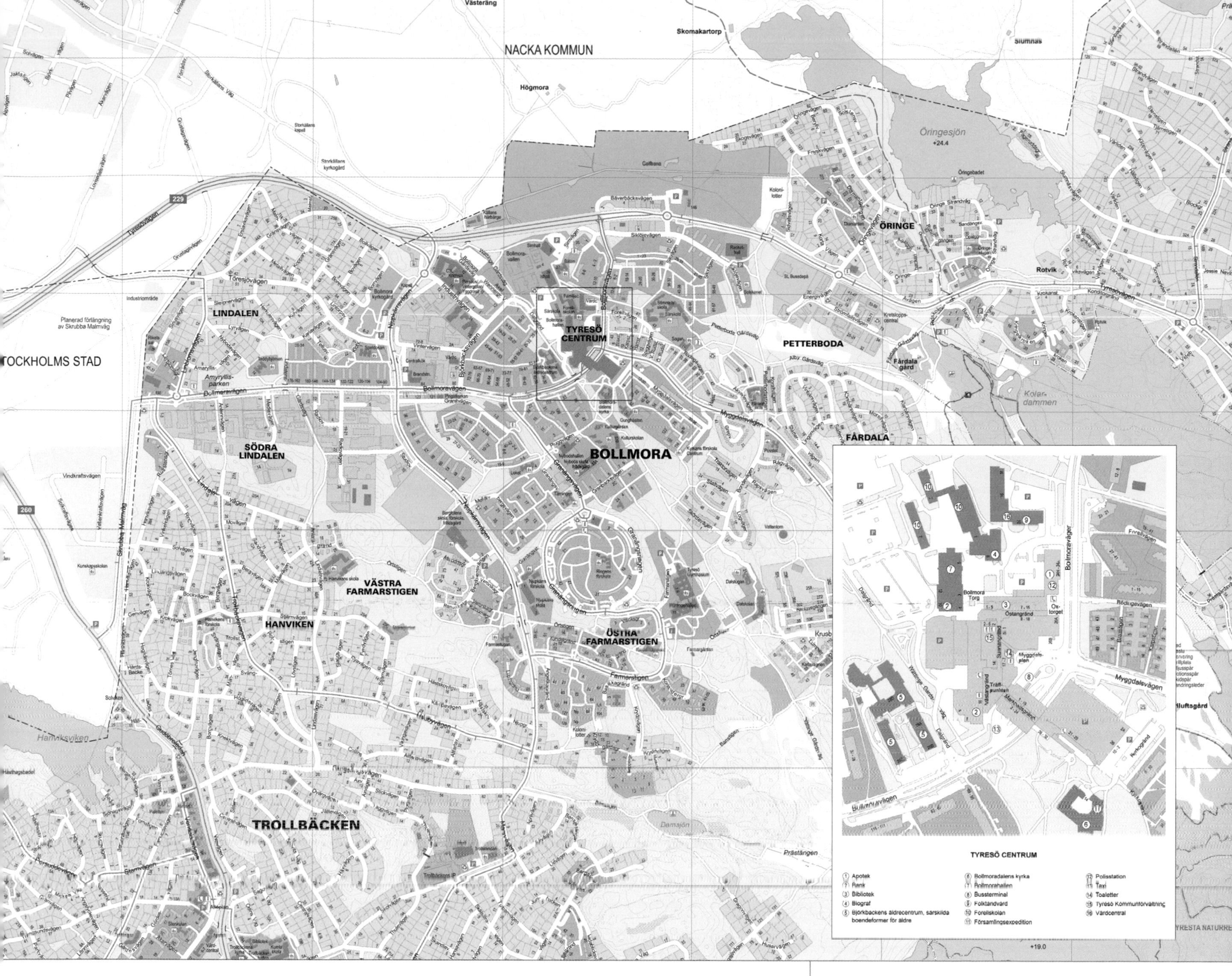

Tyresö Municipality, with about 42,000 inhabitants, is situated on the Baltic coast about 9 miles southeast of Stockholm, the Swedish capital. Cultural treasures, natural scenery, opportunities for sports and recreation, and the attractions of a metropolis are within easy reach wherever you are in Tyresö. There are frequent public transport services, and a motorway leads directly to Stockholm, whose central parts are reached in twenty minutes.

Some spots worth visiting in Tyresö include Tyresö Castle from the seventeenth century; Prinsvillan or Little Tyresö, a low-priced hostel with conference facilites; the seventeenth century Tyresö Church, one of Sweden's most popular wedding churches; and Tyresta National Park, which features rich wildlife and is close to a major city.

This tourist map is designed to reach a wide scale of users, visitors, inhabitants, and employees working with city planning. Therefore it had to have a lot of information and still be easy to read and understand.

T-Kartor Sweden AB

Kristianstad, Sweden

By Mats Persson

Contact

David Figueroa

df@t-kartor.se

Software

ArcGIS Desktop 9.2, CPS NG 2.2

Printer

Printed using DTP high-resolution digital technology

Data Source

Tyresö Municipality

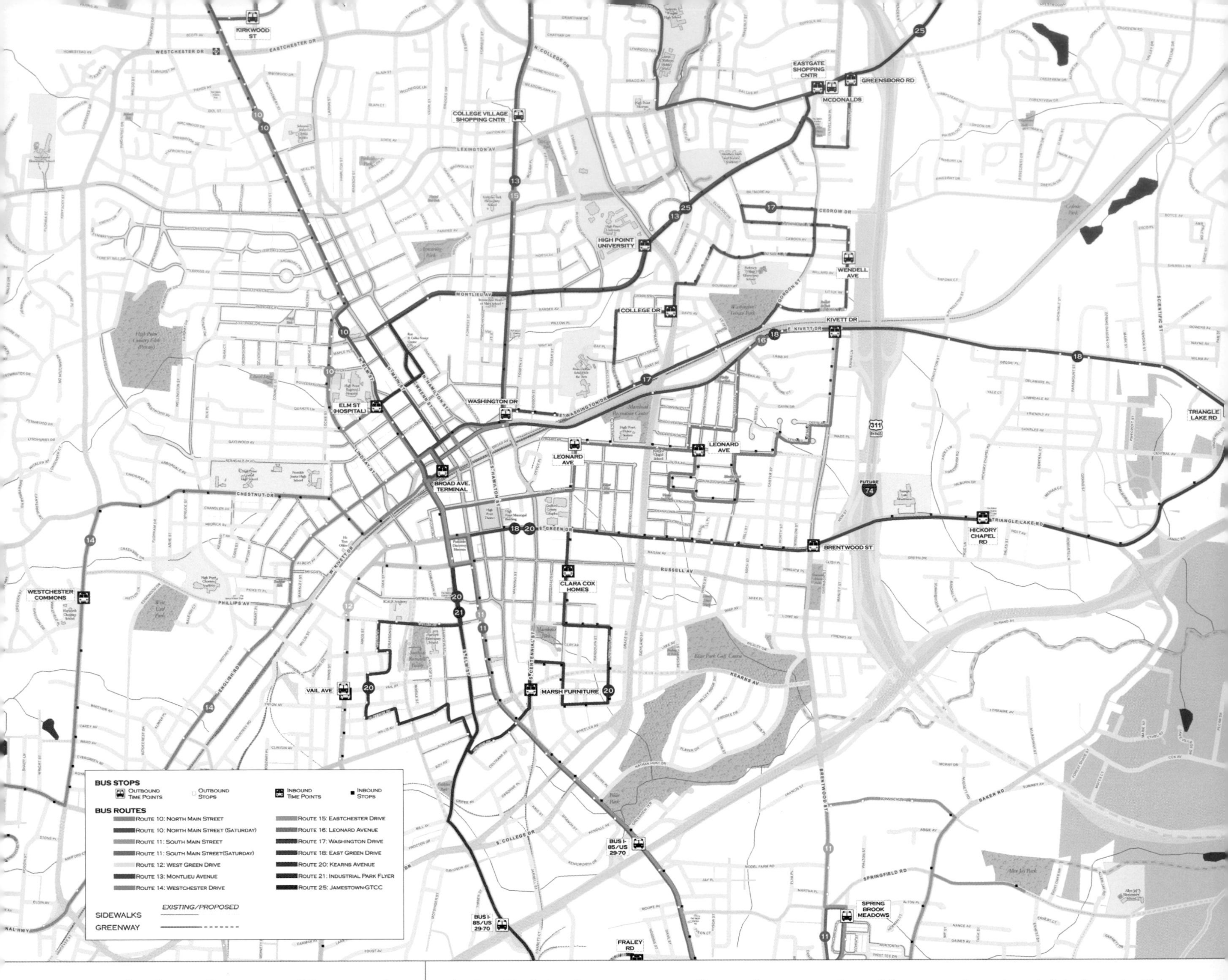

City of High Point Department of Transportation

High Point, North Carolina, USA
By Gwen Ford

Contact
Gwen Ford
gwen.ford@highpointnc.gov

Software
ArcGIS Desktop 9.2, Maplex

Printer
HP Designjet 5500ps

Data Source
City of High Point

The City of High Point bus system consists of fourteen bus routes: twelve weekday routes and two alternate Saturday routes. This series of twelve maps showing individual bus routes, stops, and time point schedules was created with a computer-aided design (CAD) software application. However, updating the maps was tedious because no attribute data was attached to the bus routes and stops.

The City of High Point Department of Transportation had already begun to convert the data to ArcGIS Desktop software. It also updated bus stop locations and attributes using recent aerial orthophotography. Creating new maps for each route was fairly simple. However, combining the routes (many of which overlap) onto one systemwide map proved more challenging without the luxuries of hiring consultants or purchasing extra ArcGIS extensions and graphic softwares.

City of High Point Bus Route Map Series

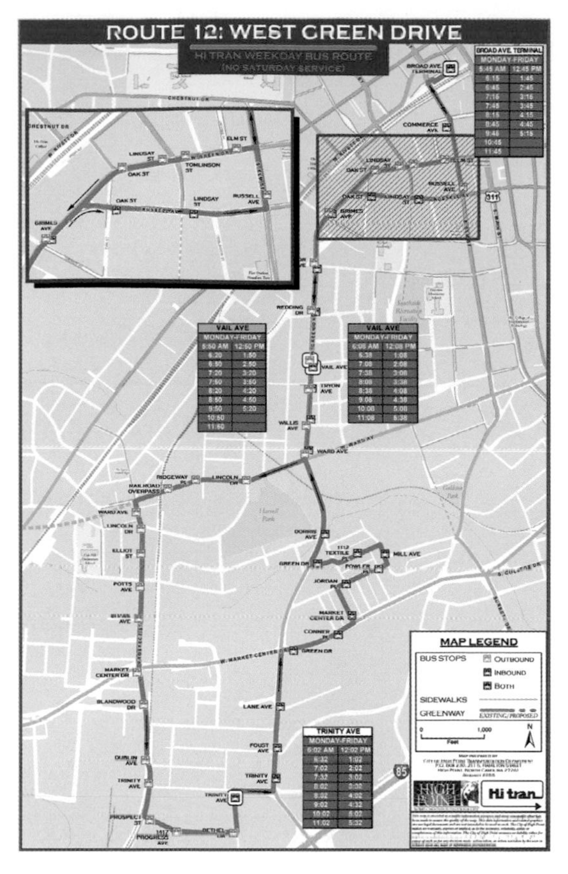

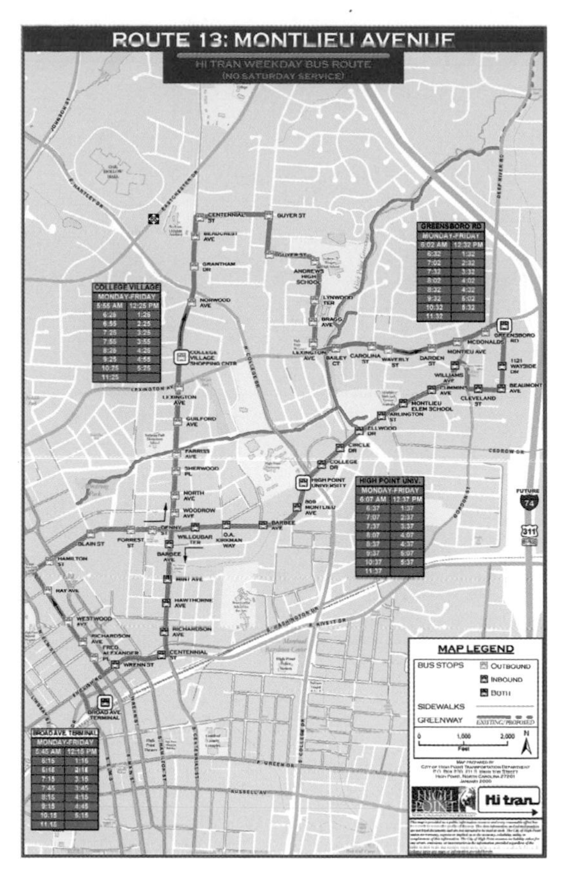

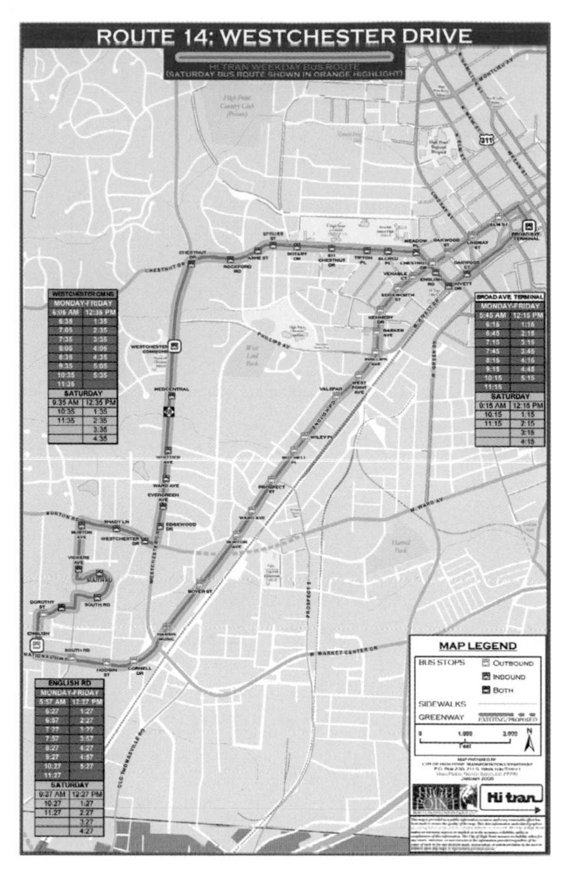

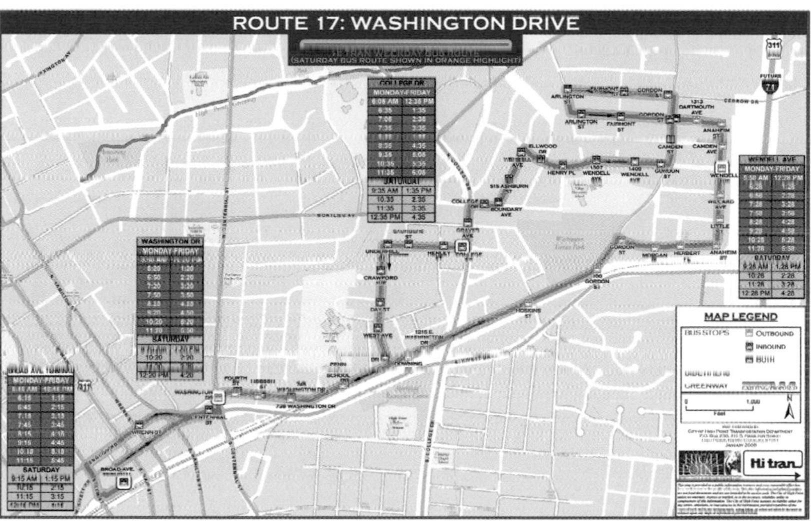

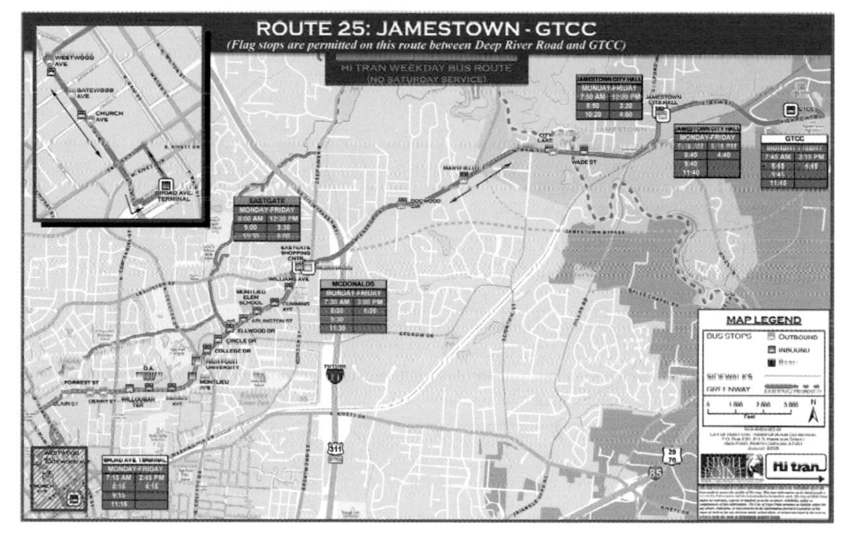

The rugged solution was to play with line directionality and offsets to show the routes that overlapped. The Maplex labeling engine saved a lot of time by automatically placing and resizing labels on the fly. The new, enhanced maps were well received by the public—so much so that the World Relief Office found them invaluable in the Refugee Care Program, which helps ease refugees' transition into the community by using the transit system to meet their daily needs. The system map was modified for World Relief Office staff to show the various locations of the refugees' homes and their destinations.

The Department of Transportation staff then generated directions for each home address to and from doctors' offices, social service offices, local shopping establishments, and the local World Relief Office. They then trained the refugees and World Relief staff how to use the transit system. The refugees now have greater mobility around their new community, and the World Relief Office has tools to help teach more newcomers how to use the transit system in the future.

Courtesy of City of High Point Department of Transportation.

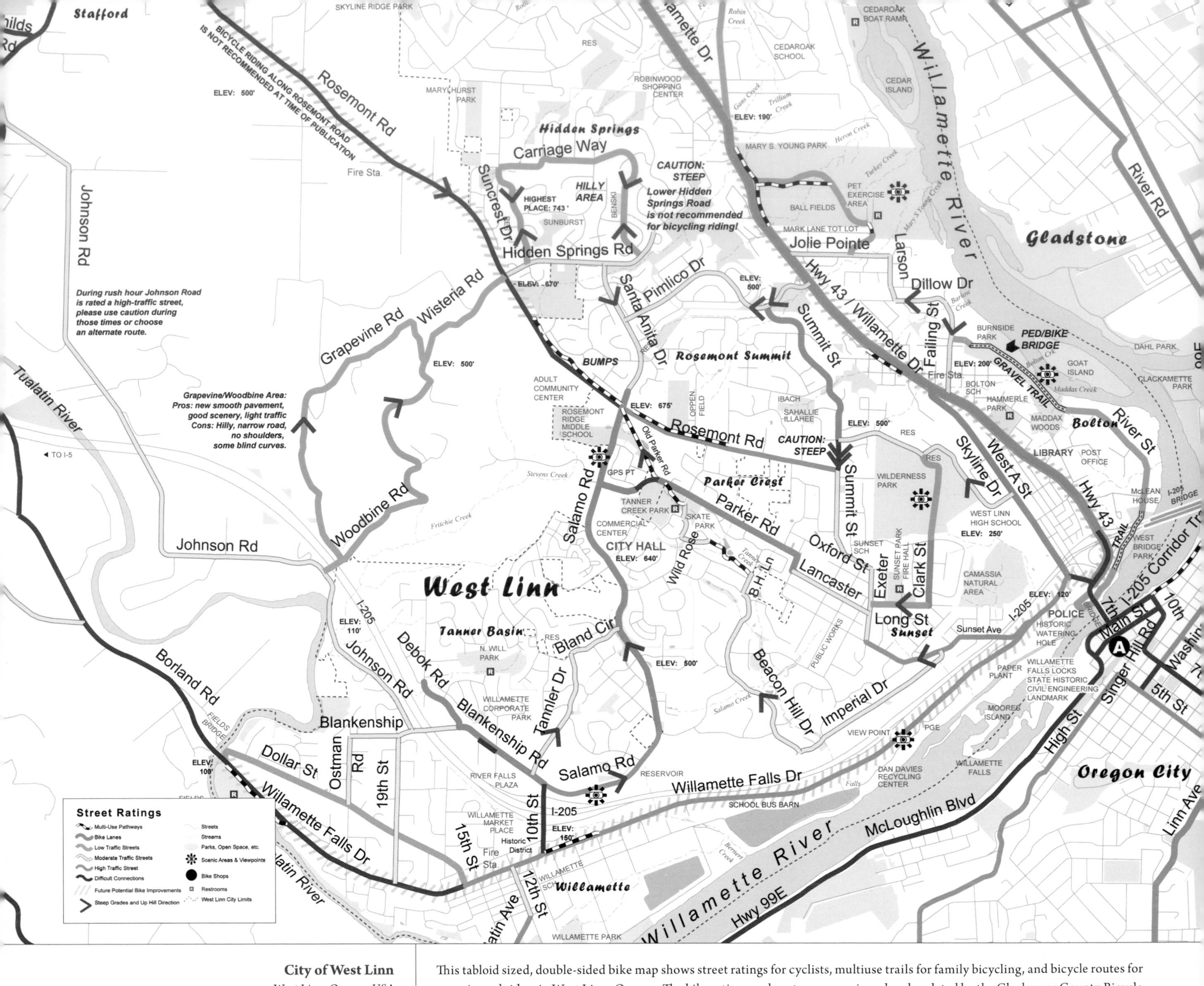

City of West Linn
West Linn, Oregon, USA
By Kathy Aha, GISP

Contact
Kathy Aha
kaha@westlinnoregon.gov

Software
ArcGIS Desktop 9.2

Printer
Xerox Phaser 7300DT laser printer

Data Source
City of West Linn GIS and Metro Regional Land Information System

This tabloid sized, double-sided bike map shows street ratings for cyclists, multiuse trails for family bicycling, and bicycle routes for experienced riders in West Linn, Oregon. The bike ratings and routes were reviewed and updated by the Clackamas County Bicycle Advisory committee members, citizens of West Linn, and city staff involved in the cycling community. Bicycle advisory committee members helped by riding paths in the city to help check the accuracy and usability of the map. The goal was to show detail for in-city riding rather than just the major routes leading in and out of the city. Scenic areas, bike shops, restrooms, and key elevations are also included on the map.

Map elements, custom symbol sets, design, and layout were created by West Linn's GIS staff using ESRI ArcGIS Desktop 9.2 software. This was the first bike map produced by the city. It was made available free to the public on the city's Web site and at City Hall in printed form at a nominal cost. The map can also be used in conjunction with the Portland metropolitan area bike map called "Bike There," produced by Metro, the regional planning agency.

City of West Linn Bike Map

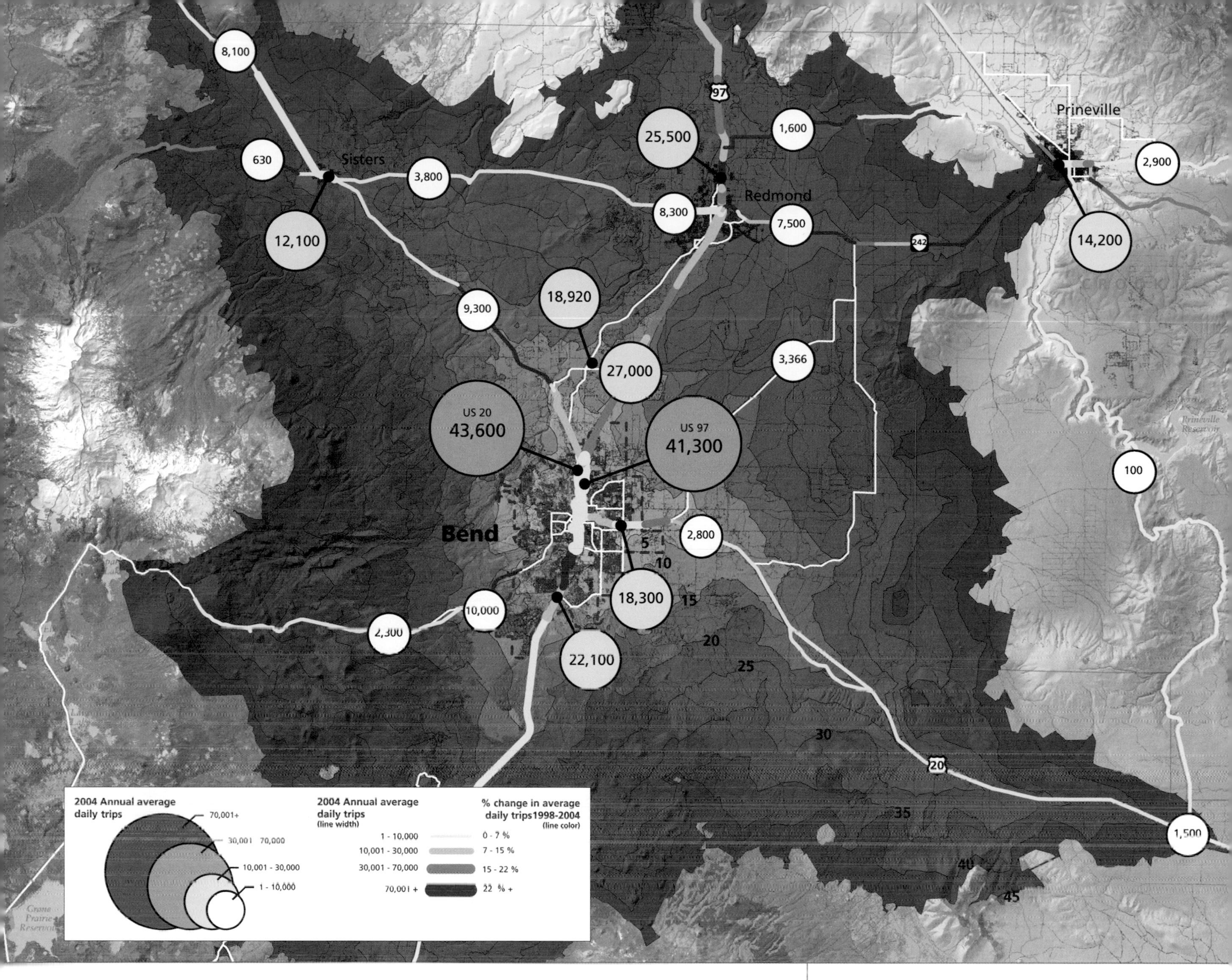

This poster examines the relationship between the Bend Metropolitan Planning Organization (MPO) and the land surrounding the Bend area. The MPO is charged with managing urban growth and sprawl while preserving farmland in the area. The map looks at travel patterns, population distribution, and land cover for the region surrounding the MPO in order to better understand the consequences and limitations of planning decisions.

Courtesy of Oregon Metro.

Oregon Metro
Portland, Oregon, USA
By Heath Brackett

Contact
Heath Brackett
heath.brackett@oregonmetro.gov

Software
ArcGIS Desktop, ArcGIS Spatial Analyst, ArcGIS Network Analyst, Microsoft Excel, Adobe Illustrator, Adobe Photoshop

Printer
HP Designjet 4500ps

Data Sources
Metro Regional Land Information System, Oregon Metropolitan Planning Organization Consortium, Oregon Department of Transportation, 2000 U.S. Census Block Groups, 2001 U.S. Geological Survey National Land Cover Data

Planning for Growth—Bend Area

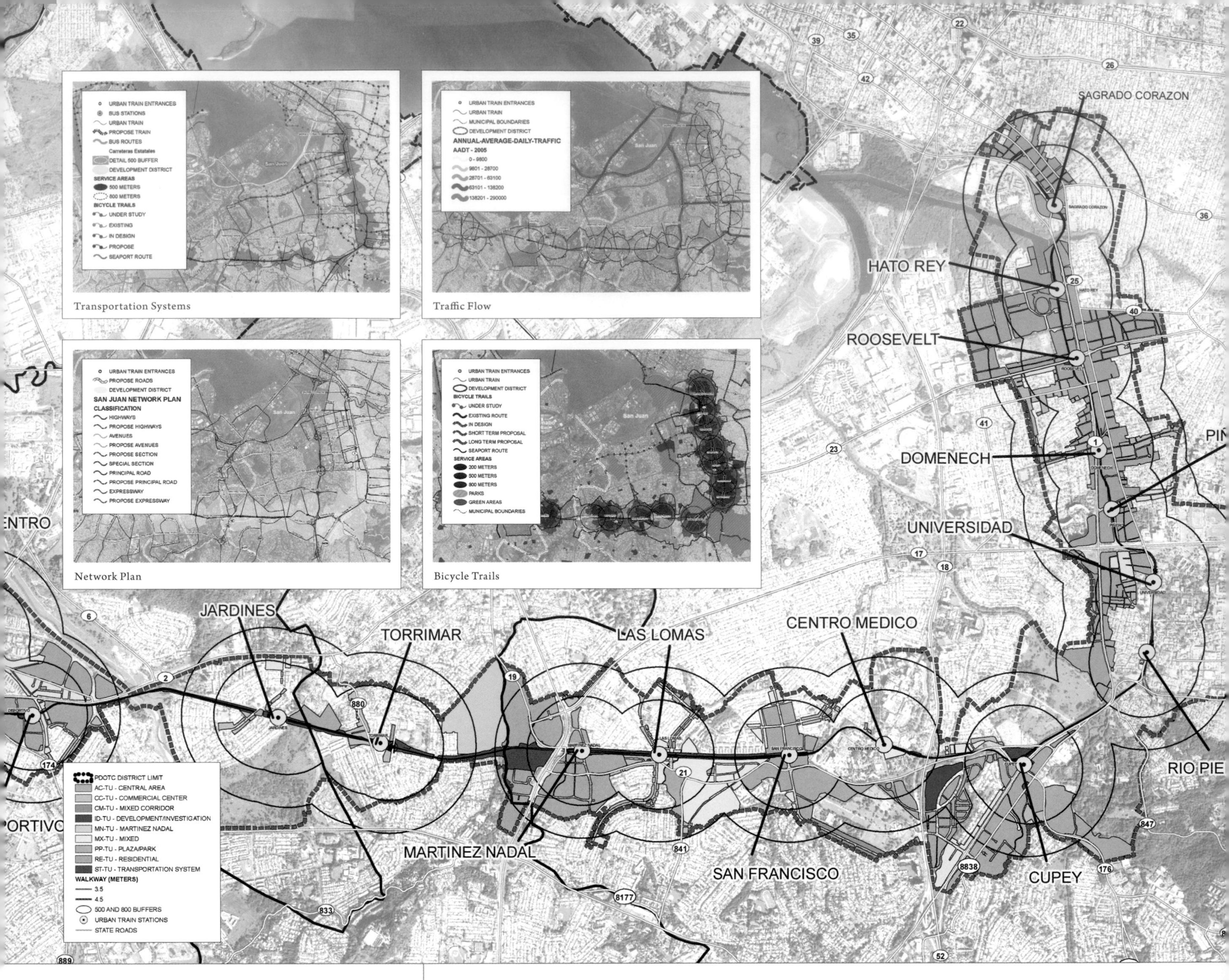

Puerto Rico Highway and Transportation Authority
San Juan, Puerto Rico, Puerto Rico, USA
By Miguel Martinez-Yordan and Luis Garcia-Pelatti

Contact
Miguel Martinez-Yordan
migmartinez@act.dtop.gov.pr

Software
ArcGIS Desktop 9.2 and 9.3, ArcGIS Network Analyst 9.2 and 9.3

Printer
HP Designjet 1055cm

Data Sources
Puerto Rico Highway and Transportation Authority, Puerto Rico Planning Board, and Center for the Collection of Municipal Revenues

This is a representation of the Transport Oriented Development Plan by the Puerto Rico Highway and Transportation Authority (PRHTA). It is a transportation and land-use plan to increase density in terms of housing, business, and public and private services in areas surrounding the main entrances of the Puerto Rico Urban Train.

As part of this plan, a 500-meter (547-yard) buffer was outlined from each train station, and special zoning considerations were taken into account in order to redefine the public needs. The plan covered the train corridor's effects on housing, security, business operations, and public agencies. The PRHTA, operating under the Department of Transportation and Public Works, was in charge of developing this plan and presenting it to the Puerto Rico Planning Board for approval and implementation.

ArcGIS Desktop 9.2 and 9.3 software was used to create and manage the data. Cadastral records and orthophotogrammetry were used to set the base layers for the analysis. Actual zoning districts were analyzed to meet the needs of the plan, and the original urban train "as builds" was overlaid to outline the new zoning districts for the final plan. Assisting in this project were Rebecca de la Cruz of the Puerto Rico Planning Board and Lisandra Benitez of the Center for the Collection of Municipal Revenues.

Courtesy of the Puerto Rico Highway and Transportation Authority.

Transport Oriented Development Plan

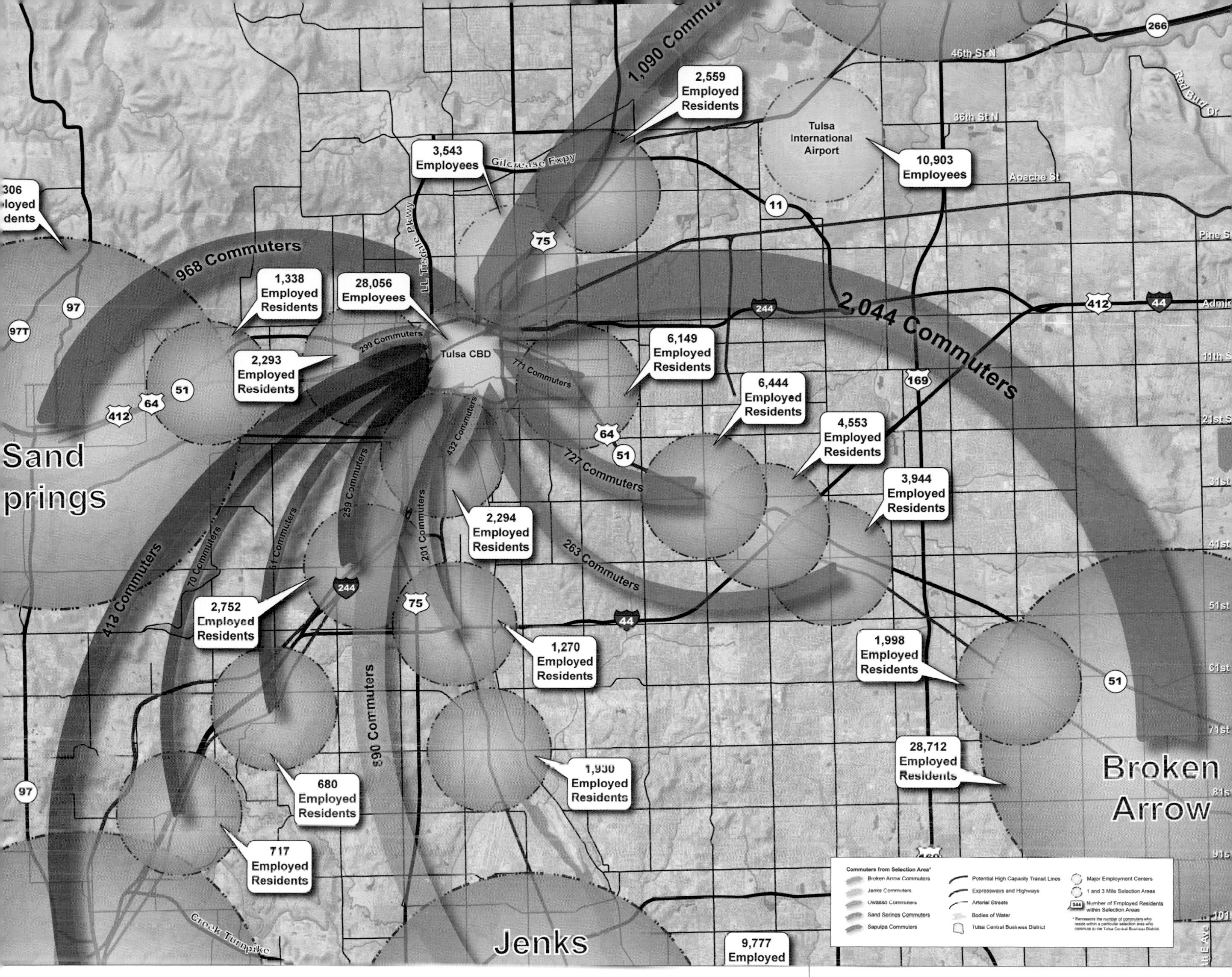

In analyzing the potential for passenger rail service, INCOG, the metropolitan planning organization for the Tulsa region, compiled data regarding commuter patterns and employment. This map shows the number of employed residents within suburban communities (3-mile radius) and potential Transit Oriented Development (1-mile radius) "selection areas" along existing rail corridors with potential as high-capacity transit lines. In addition, this map illustrates the number of commuters residing within the selection areas who commute to the Central Business District.

Courtesy of INCOG.

INCOG
Tulsa, Oklahoma, USA
By Ty Simmons

Contact
Ty Simmons
tsimmons@incog.org

Software
ArcGIS Desktop 9.0, ArcSDE 9.2, and Adobe Illustrator

Printer
HP Designjet 5500ps

Data Sources
U.S. Census Bureau, Oklahoma Department of Commerce, Oklahoma Employment Security Commission, Internal Revenue Service, and U.S. Department of Labor

Tulsa Existing Commuter Shed Statistics

All Jet Traffic Flight Paths, Red: Departures, Blue: Arrivals

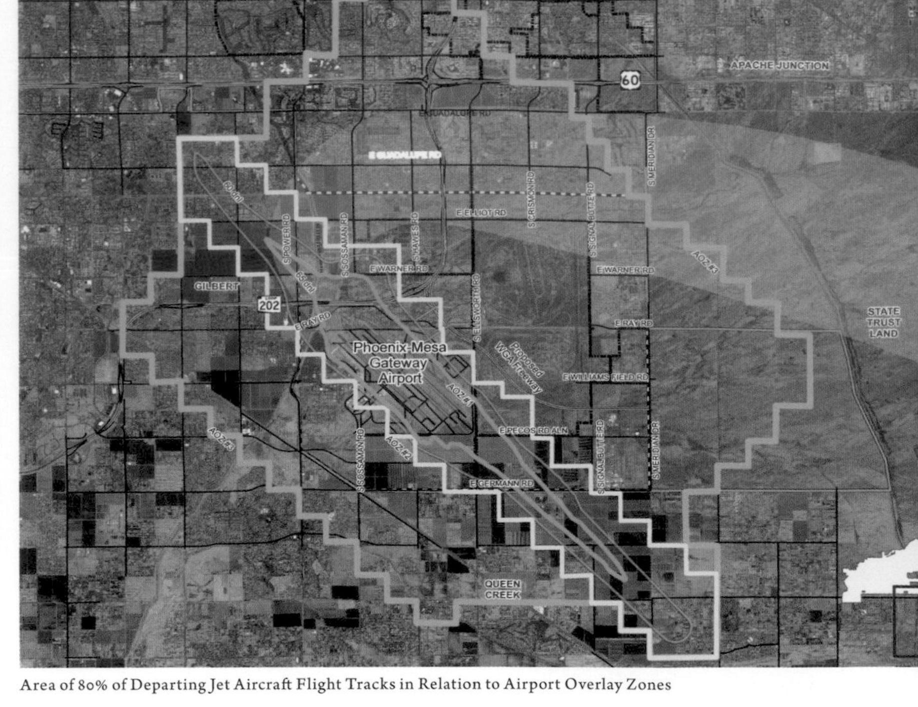

Area of 80% of Departing Jet Aircraft Flight Tracks in Relation to Airport Overlay Zones

Density of Departing Flight Paths, All Jet Traffic

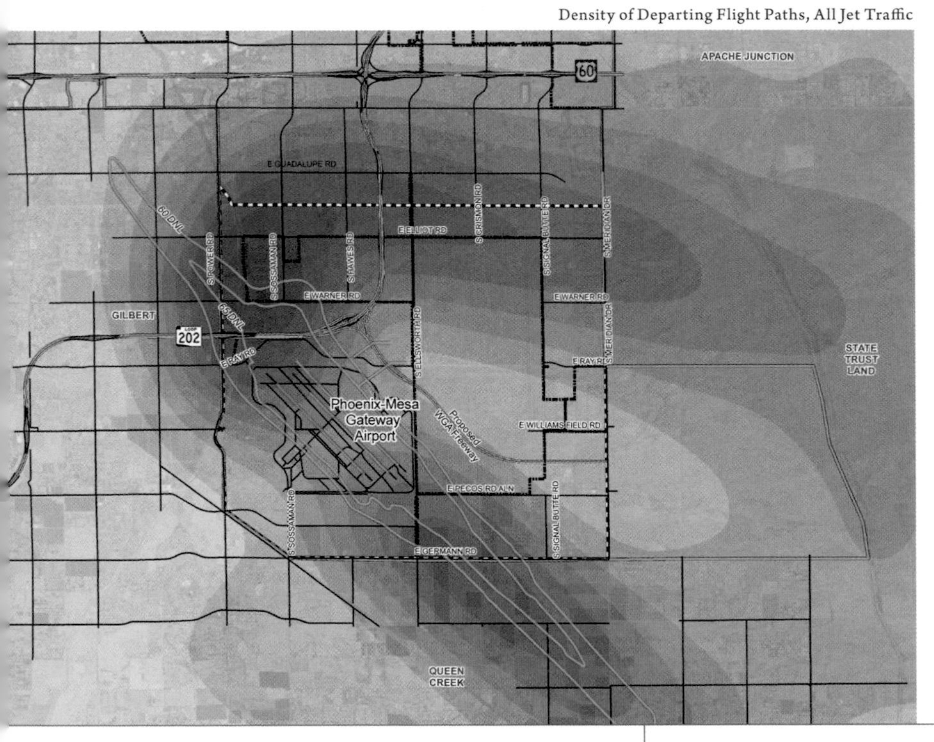

Density of Departing Flight Paths, Helicopter Traffic

City of Mesa
Mesa, Arizona, USA
By Tony Bianchi and Cory Whittaker

Contact
Cory Whittaker
cory.whittaker@mesaaz.gov

Software
ArcGIS Desktop 9.2, ArcGIS Spatial Analyst, Adobe Illustrator CS3

Printer
HP Designjet 1055cm

Data Source
City of Mesa

Phoenix-Mesa Gateway Airport (formerly Williams Gateway Airport) is now the main commercial reliever airport to Sky Harbor Airport in the Phoenix Metro area. Proper planning of the surrounding area can ensure the continued growth of airport operations and be the catalyst for further economic development for the City of Mesa. By analyzing the flight patterns using ArcGIS Spatial Analyst, appropriate land uses can be identified where aircraft operations would have the least impact.

Courtesy of City of Mesa, Arizona.

Flight Path Analysis at Phoenix-Mesa Gateway Airport

Federal Aviation Regulations (FAR) Part 77: Analysis of Acceptable Structure Height for the Airport Vicinity

This Swath Represents the Area for 80% of Departing Jet Aircraft Based on Flight Track Data

This 80% Swath in 3D Shows Average Altitude and Rate of Climb for Departing Aircraft

Noise Footprint Based on Average Flight Path of Departing MD-80 Jet Aircraft with Outer Ring at 60db

Using ArcGIS 3D Analyst in conjunction with Google Earth, Federal Aviation Administration regulations regarding allowable building heights can be better displayed for modeling purposes. In addition, with the assistance of ATAC Corporation's tools, aircraft noise can be modeled to show the extent of areas most prone to aircraft over flight for both existing and proposed development.

Courtesy of City of Mesa, Arizona.

City of Mesa
Mesa, Arizona, USA
By Tony Bianchi and Cory Whittaker

Contact
Cory Whittaker
cory.whittaker@mesaaz.gov

Software
ArcGIS Desktop 9.2, ArcGIS 3D Analyst, Google Earth

Printer
HP Designjet 1055cm

Data Sources
City of Mesa, Arizona State University Decision Theater, ATAC Corporation

Sound and Safety Analysis at Phoenix-Mesa Gateway Airport

City of Santa Monica

Santa Monica, California, USA

By Sharon Wong and Michael A. Carson

Contact

Michael A. Carson

michael.carson@smgov.net

Software

ArcGIS Desktop 9.2, Adobe Illustrator, MAPublisher

Printer

HP Designjet 5500 ps

Data Sources

Santa Monica GIS, Santa Monica Transportation, Santa Monica Big Blue Bus, BikeSantaMonica.org

This bike map was created to increase bike ridership and to inform the public about bike safety in Santa Monica. This map shows streets that have designated bike lanes, bike paths, and bike routes throughout the city. Included in the map are bike-rider amenities such as bike parking, restrooms, bike shops, and repair service locations. The bike map was designed to be folded into a convenient size for travel.

Bike Santa Monica

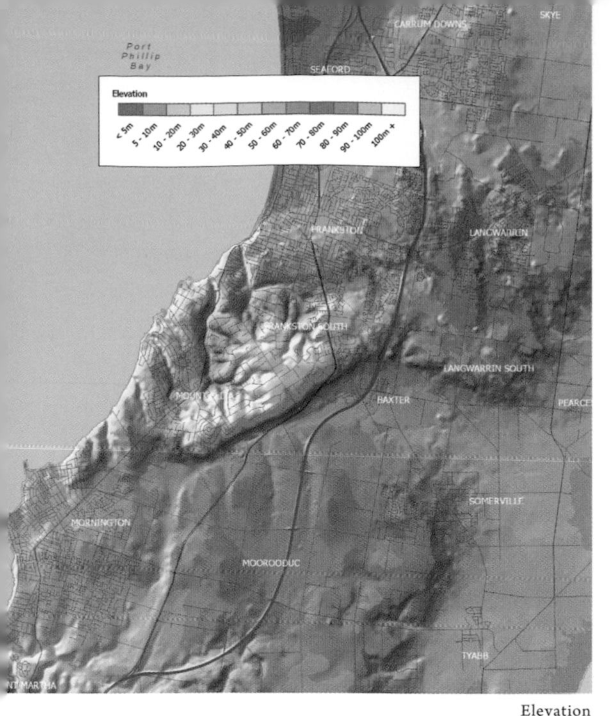

Elevation

Slope Analysis

Land-Use/Zoning

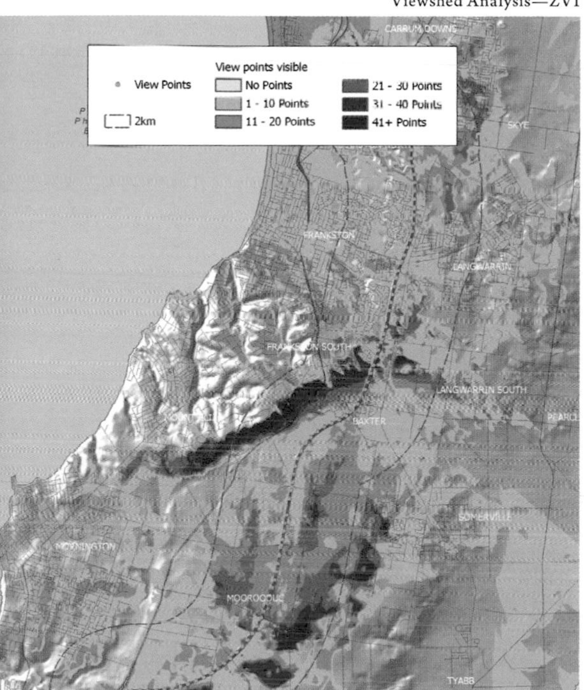

Viewshed Analysis—ZVI

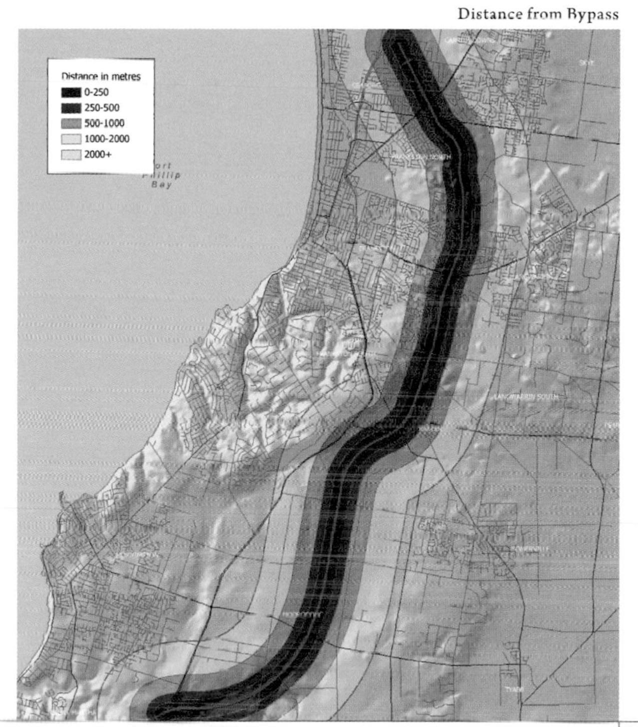

Distance from Bypass

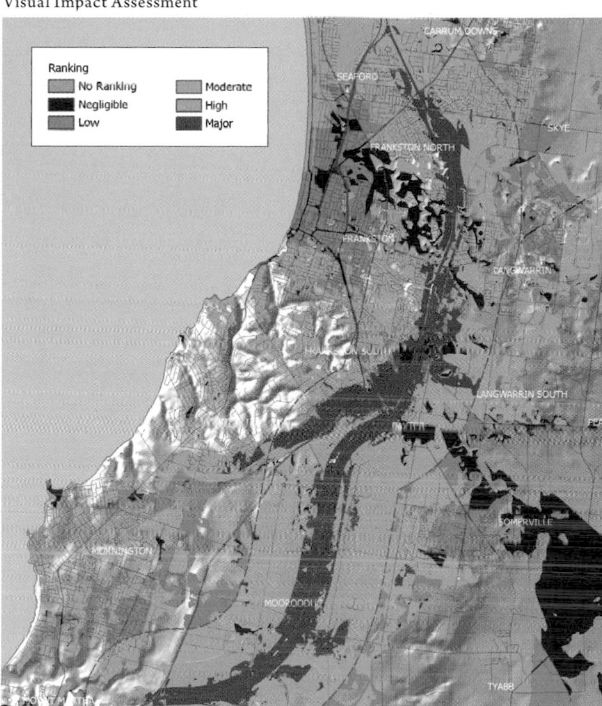

Visual Impact Assessment

EDAW was engaged by Maunsell AECOM on behalf of the Southern and Eastern Integrated Transport Authority (SEITA) to undertake an environmental assessment of landscape and visual impacts that would result from the proposed Frankston Bypass. SEITA proposed the development of a freeway to bypass the bayside city of Frankston 30 kilometers (18.6 miles) south of Melbourne.

The primary objective of the Frankston Bypass is to achieve a continuous and balanced road network into the future with sufficient road system capacity in the Frankston-Mornington Peninsular corridor to meet the likely road travel demands resulting from Melbourne 2030—Planning for Sustainable Growth.

EDAW used ArcGIS software throughout the project for the Visual Impact Analysis and report mapping.

Courtesy of EDAW and AECOM.

EDAW AECOM

Melbourne, Victoria, Australia

By Geoff Williams

Contact

Geoff Williams

geoff.williams@edaw.com

Software

ArcGIS Desktop 9.2, Adobe CS2

Printer

HP Designjet Z6100ps

Data Sources

Vicmap www.land.vic.gov.au, Maunsell AECOM

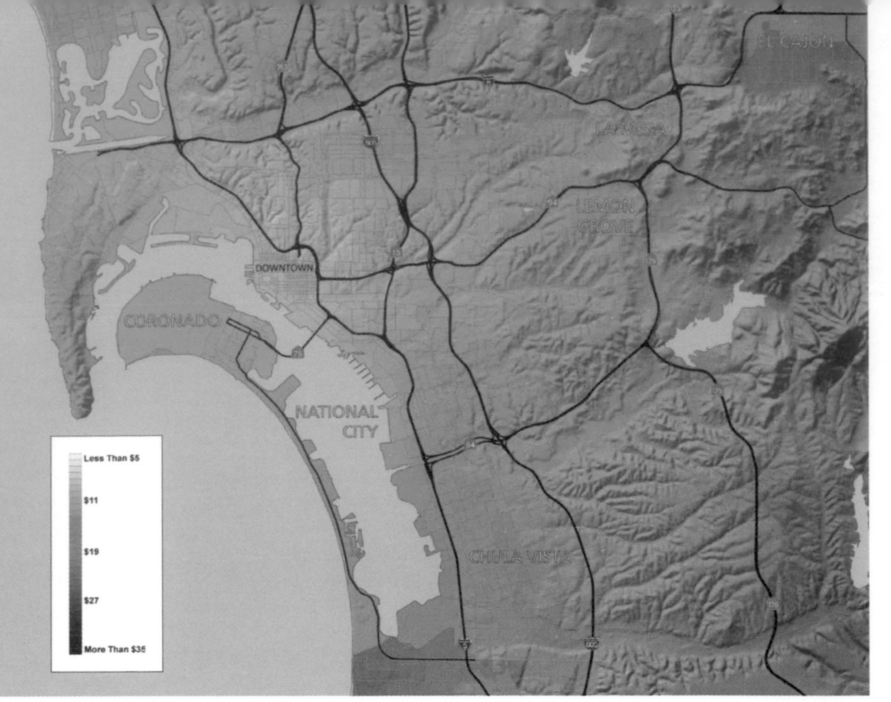

2030 Single Auto Morning Commute Cost

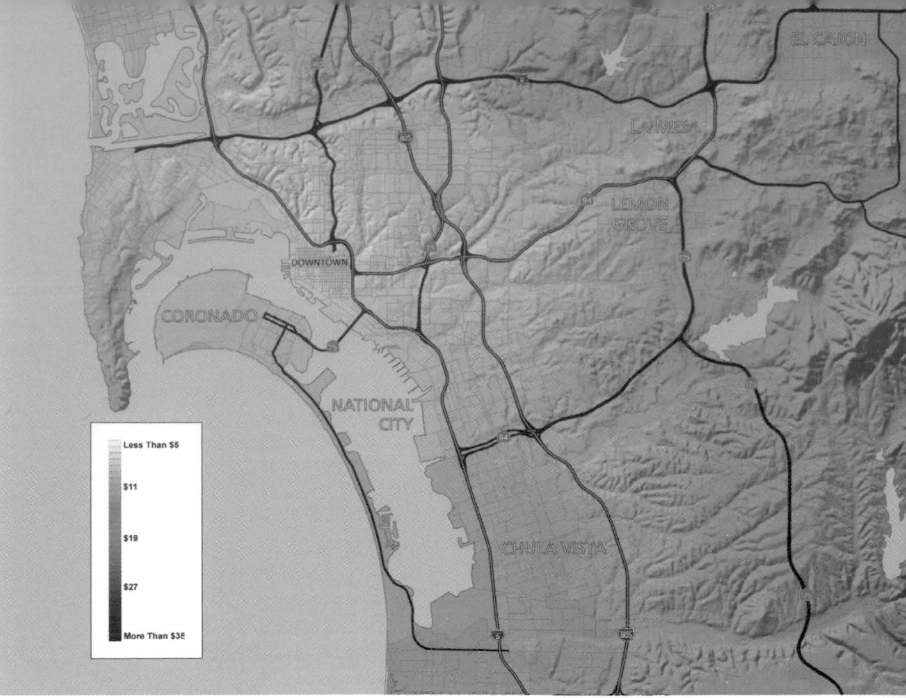

2030 Carpool Auto Morning Commute Cost

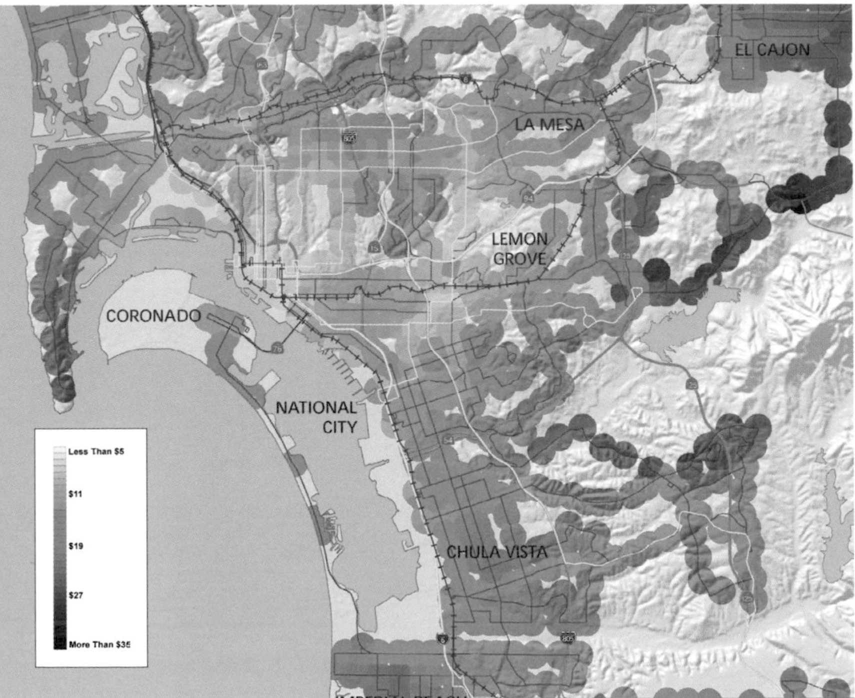

2030 Transit Morning Commute Cost

2030 Transit Value Areas

San Diego Association of Governments
San Diego, California, USA
By Joaquin S. Ortega

Contact
Joaquin Ortega
jor@sandag.org

Software
ArcGIS Desktop 9.2

Printer
HP Designjet 800ps

Data Sources
San Diego Association of Governments, SanGIS, and Aerials Express

These maps display the forecasted cost of a morning trip to downtown San Diego. The first three panels display costs for three modes: single occupant auto, high occupancy auto, and walk-access transit. Auto trip cost is calculated by using forecasted trip length and trip time data from the transportation model multiplied by cost factors for auto maintenance, tire degradation, fuel consumption, toll road use, and value of time.

A fixed fee for parking is also added to all trips ending in the central business district which included a carpool incentive discount for high-occupancy vehicle (HOV) trips. Free use of planned managed lane facilities, along with trip time savings from use of dedicated HOV facilities gives carpools an added cost savings over single auto travel. Transit trip cost is calculated by using forecasted transit trip length and trip time data (including access and transfers) from the transportation model which is multiplied by cost factors for transit fare and value of time.

The fourth panel shows transit value areas by comparing the cost of single occupant auto travel to transit. The green areas shown on this map are neighborhoods in the San Diego region where the total cost of using transit is 15 percent less than driving solo in an automobile. Eighteen cities and the county government comprise SANDAG, the San Diego Association of Governments, which serves as a forum for regional decision making.

Courtesy of Joaquin Ortega, San Diego Association of Governments.

Downtown San Diego—2030 Commute Cost Analysis

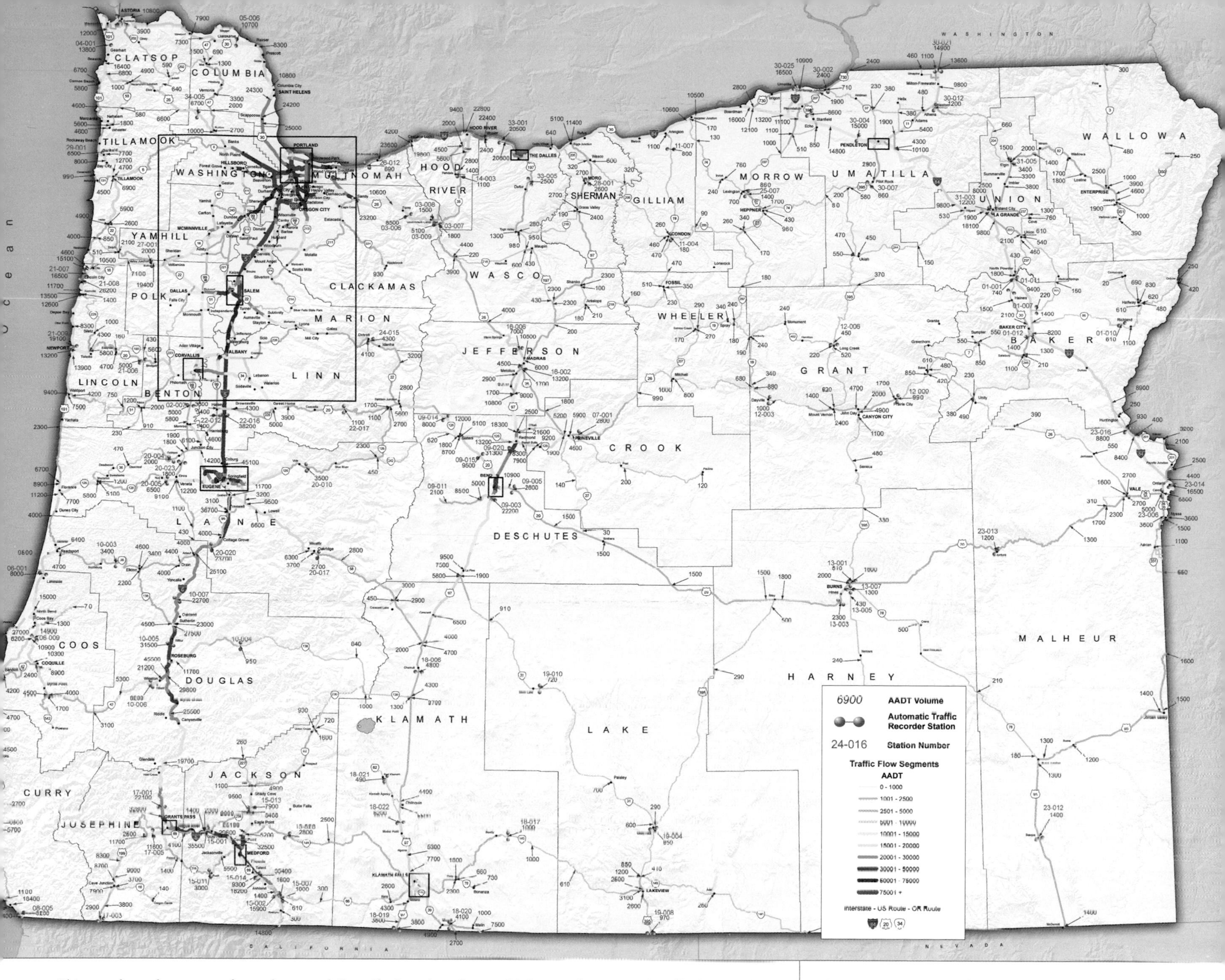

This map shows the patterns of annual average daily traffic throughout Oregon. With just under 9,000 miles of highway and over 20 billion vehicle miles traveled annually, managing these systems is a monumental effort. The Traffic Flow Map is an annual publication that the Oregon Department of Transportation (ODOT) sends out to government agencies, private companies, and the public. Agencies such as the Oregon State Police, marketing firms, real estate companies, engineering firms, and the Tourism Commission all have unique uses for this information and are annual customers of the Traffic Flow Map.

This map has been produced for over fifty years and has evolved along with advancing technologies in cartography. This map was converted to GIS in 2004, and the current version is stored in ODOT's enterprise geodatabase with the cartography being done using ArcGIS software. Traffic volume and classification data is collected by 150 automatic traffic recorders and around 5,000 count sites that use hose tubes draped across the roadway. The data updates are now an automated workflow that can be completed in a few minutes, using Feature Manipulation Engine, whereas the previous versions would take a few weeks or longer.

Courtesy of Oregon Department of Transportation, Transportation Systems Monitoring.

Oregon Department of Transportation
Salem, Oregon, USA
By Ryan R. Johnson

Contact
Ryan R. Johnson
Ryan.R.Johnson@odot.state.or.us

Software
ArcGIS Desktop 9.2, Adobe Photoshop CS2

Printer
HP Designjet 5500 ps

Data Source
Oregon Department of Transportation

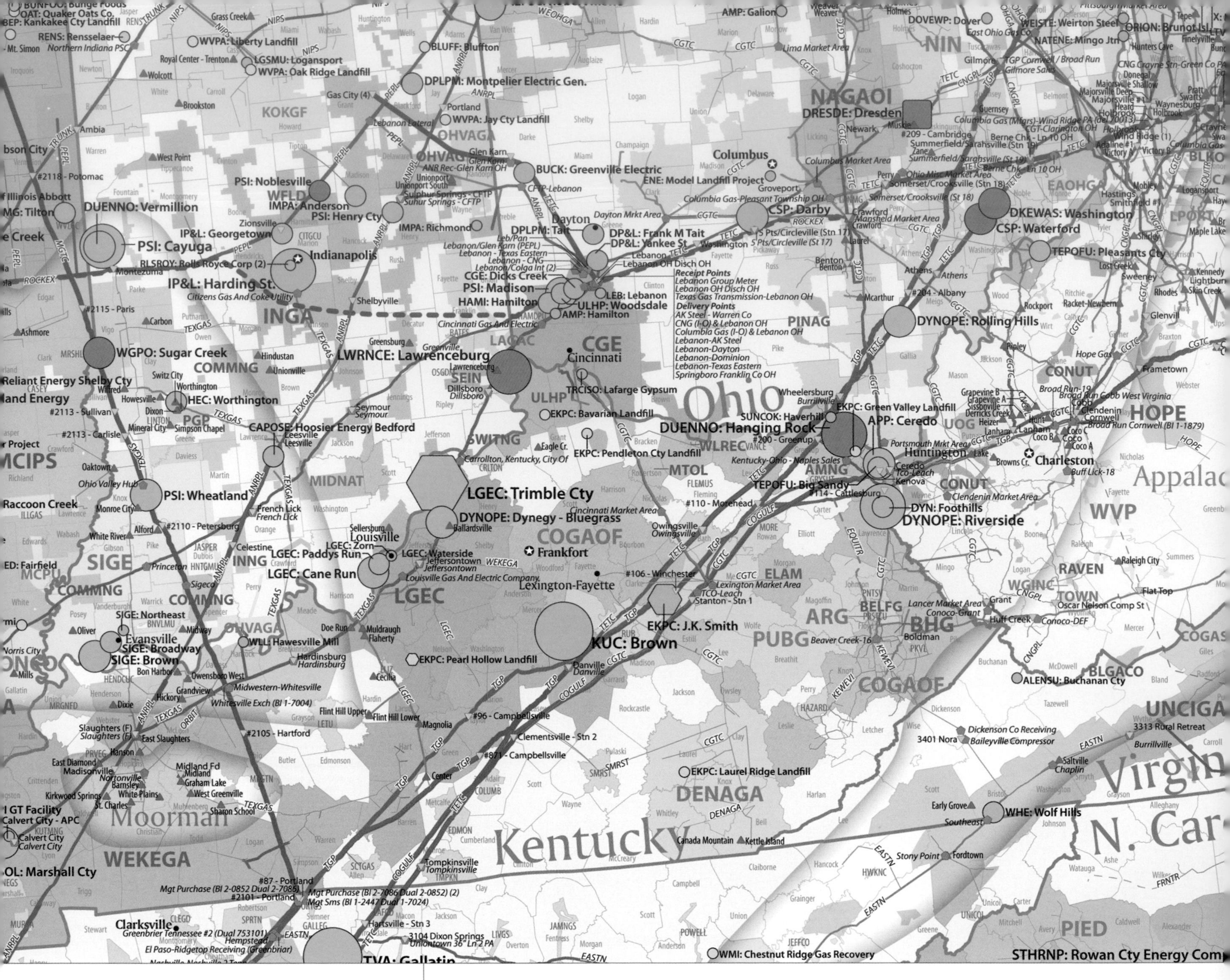

Platts (McGraw-Hill)
Denver, Colorado, USA
By Claude Frank and Erin LeFevre

Contact
Claude Frank
claude_frank@platts.com

Software
ArcGIS Desktop, Adobe Illustrator

Printer
HP Designjet 5000ps

Data Source
Platts POWERmap

The North American Natural Gas System wall map provides an in-depth look at the supply, transportation, and demand aspects of the evolving natural gas marketplace. The map's objective is to enable the viewer to decipher broad patterns in the industry while exploring detailed information about individual plants, compressor stations, pipelines, and other key infrastructure components.

Customers span the industry from business leaders to GIS analysts. The map is used in their daily decision making process to formulate strategy, for construction planning, and for generation and transmission management.

Platts, a division of the McGraw-Hill Companies, is a leading global provider of energy and metals information. Platts serves customers across the oil, natural gas, electricity, nuclear power, coal, petrochemical, and metals markets of more than 150 countries. Platts' geospatial data, map products, news, pricing, analytical services, and conferences help markets operate with transparency and efficiency.

The North American Natural Gas System

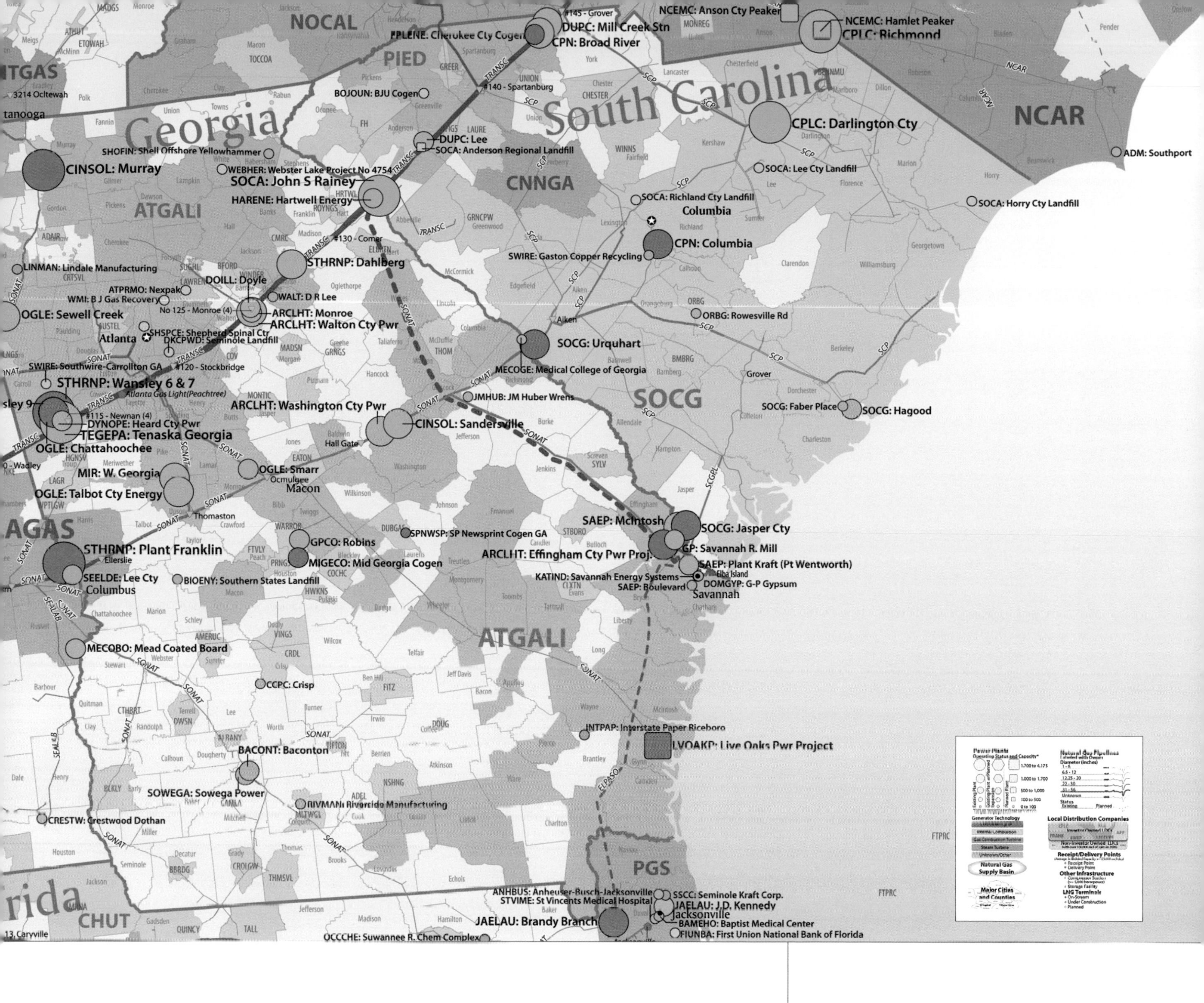
NOCAL
PIED
Georgia
South Carolina
NCAR
CNNGA
ATGALI
SOCG
PGS
CHUT
AGAS
TGAS
NCEMC: Anson Cty Peaker
NCEMC: Hamlet Peaker
CPLC: Richmond
#145 - Grover
DUPC: Mill Creek Stn
CPN: Broad River
FPLENE: Cherokee Cty Cogen
#140 - Spartanburg
BOJOUN: BJU Cogen
CPLC: Darlington Cty
ADM: Southport
DUPC: Lee
SOCA: Anderson Regional Landfill
SHOFIN: Shell Offshore Yellowhammer
WEBHER: Webster Lake Project No 4754
CINSOL: Murray
SOCA: John S Rainey
HARENE: Hartwell Energy
SOCA: Lee Cty Landfill
SOCA: Richland Cty Landfill
Columbia
SOCA: Horry Cty Landfill
#130 - Comer
CPN: Columbia
SWIRE: Gaston Copper Recycling
STHRNP: Dahlberg
LINMAN: Lindale Manufacturing
DOILL: Doyle
ATPRMO: Nexpak
WMI: B J Gas Recovery
WALT: D R Lee
No 125 - Monroe (4)
OGLE: Sewell Creek
ARCLHT: Monroe
ARCLHT: Walton Cty Pwr
ORBG: Rowesville Rd
SHSPCE: Shepherd Spinal Ctr
Atlanta
DKCPWD: Seminole Landfill
SOCG: Urquhart
SWIRE: Southwire-Carrollton GA
#120 - Stockbridge
MECOGE: Medical College of Georgia
STHRNP: Wansley 6 & 7
Atlanta Gas Light(Peachtree)
JMHUB: JM Huber Wrens
ARCLHT: Washington Cty Pwr
SOCG: Faber Place
SOCG: Hagood
#115 - Newnan (4)
DYNOPE: Heard Cty Pwr
CINSOL: Sandersville
TEGEPA: Tenaska Georgia
Hall Gate
OGLE: Chattahoochee
OGLE: Smarr
MIR: W. Georgia
Macon
OGLE: Talbot Cty Energy
Thomaston
SAEP: McIntosh
SOCG: Jasper Cty
SPNWSP: SP Newsprint Cogen GA
GPCO: Robins
GP: Savannah R. Mill
STHRNP: Plant Franklin
Ellerslie
ARCLHT: Effingham Cty Pwr Proj.
MIGECO: Mid Georgia Cogen
SAEP: Plant Kraft (Pt Wentworth)
KATIND: Savannah Energy Systems
Elba Island
SEELDE: Lee Cty
BIOENY: Southern States Landfill
DOMGYP: G-P Gypsum
Columbus
SAEP: Boulevard
Savannah
MECOBO: Mead Coated Board
CCPC: Crisp
INTPAP: Interstate Paper Riceboro
LVOAKP: Live Oaks Pwr Project
BACONT: Baconton
SOWEGA: Sowega Power
RIVMAN: Riverside Manufacturing
CRESTW: Crestwood Dothan
FTPRC
ANHBUS: Anheuser-Busch-Jacksonville
STVIME: St Vincents Medical Hospital
SSCC: Seminole Kraft Corp.
JAELAU: J.D. Kennedy
JAELAU: Brandy Branch
Jacksonville
BAMEHO: Baptist Medical Center
FIUNBA: First Union National Bank of Florida
OCCCHE: Suwannee R. Chem Complex
TRANSC
SCP
SONAT
ELPASO
SCGPL
Local Distribution Companies
Receipt/Delivery Points
Other Infrastructure
LNG Terminals
Natural Gas Supply Basin
Major Cities and Counties

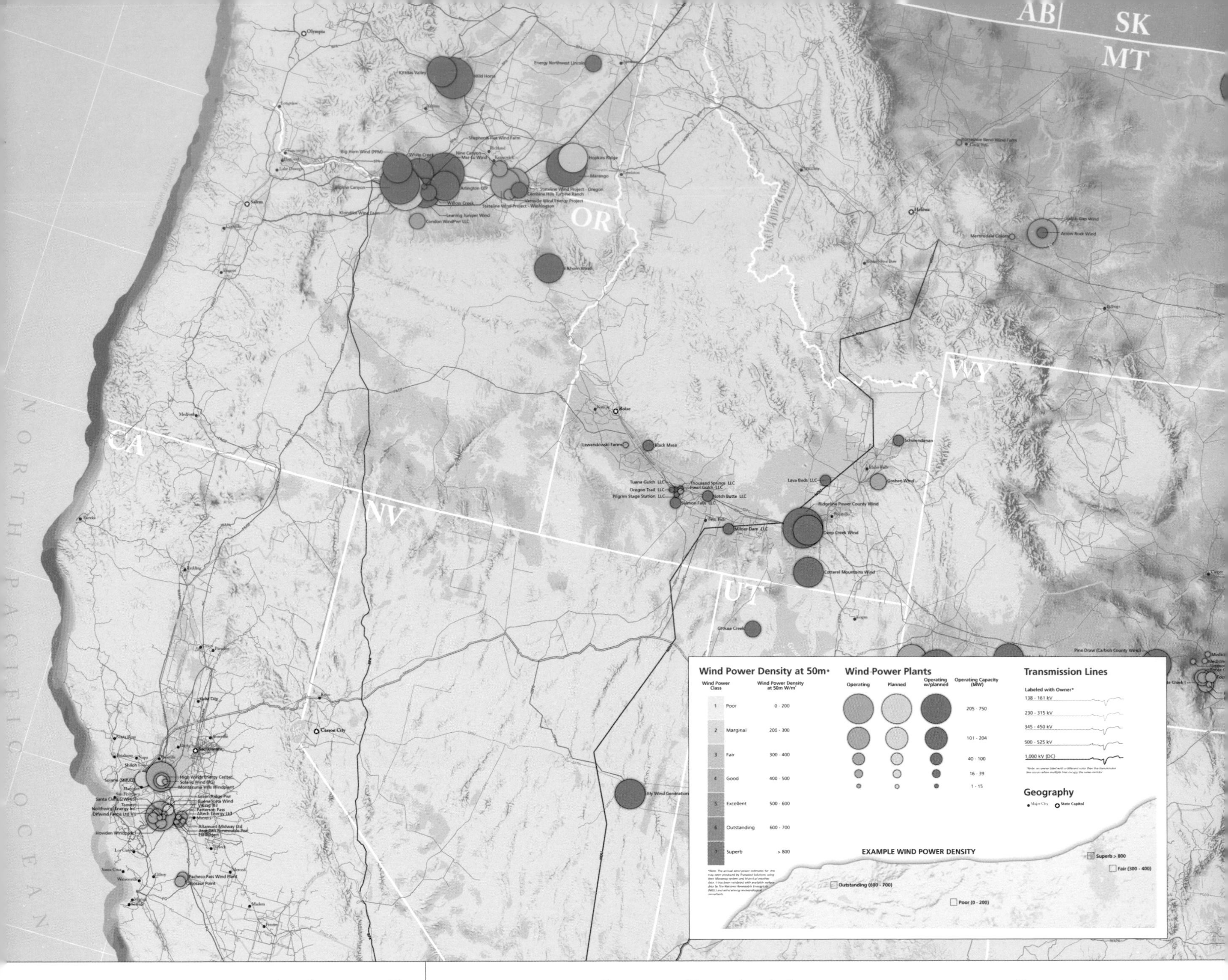

Platts (McGraw-Hill)
Denver, Colorado, USA
By Erin LeFevre and Claude Frank

Contact
Erin LeFevre
erin_lefevre@platts.com

Software
ArcGIS Desktop, Adobe Illustrator

Printer
HP Designjet 5000ps

Data Source
Platts POWERmap

In less than thirty years, wind has progressed from a relatively obscure power source into an exciting, rapidly growing power industry. New methods of predicting suitable locations for wind infrastructure have become exceptionally accurate, making wind power more economical, reliable, and scalable.

By presenting current wind generation, transmission options, and new wind generation siting opportunities, Platts' clients use the Wind Resources of the Western United States wall map to obtain insight into the present and future of wind power in the western United States.

Platts, a division of the McGraw-Hill Companies, is a leading global provider of energy and metals information. Platts serves customers across the oil, natural gas, electricity, nuclear power, coal, petrochemical, and metals markets of more than 150 countries. Platts' geospatial data, map products, news, pricing, analytical services, and conferences help markets operate with transparency and efficiency.

Wind Resources of the Western United States, 2007–2008 Edition

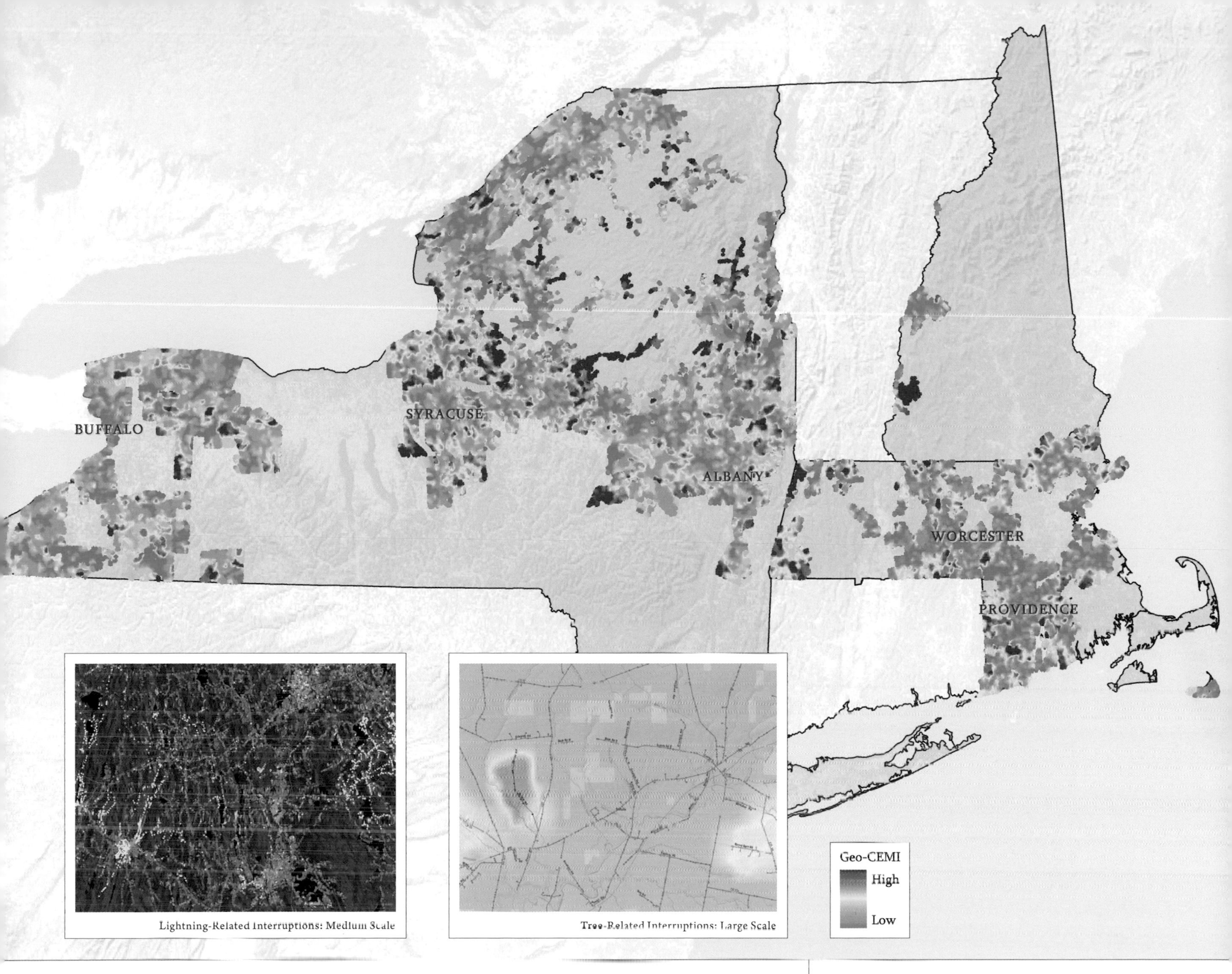

Traditional reliability measures for electric utilities are based on average interruption frequency, usually for entire circuits, but also for boundaries such as towns or operating districts. For example, the CEMI metric (Customers Experiencing Multiple Interruptions) is applied to a circuit or boundary and expressed as a percentage.

These generalized statistics have been effective through the years, but advances in technology are now able to capture each individual customer's experience. Combined with the analysis and visualization power of GIS, this new wealth of information provides insight beyond the averages, exposing geographic pockets of poor performance that were previously masked by conventional metrics.

Derived from CEMI, Geo-CEMI geographically weights the distribution of electricity reliability. With customer-level interruption and location data at its source, this new way of analyzing reliability has great potential for optimizing asset management. By identifying or projecting trouble spots and then taking appropriate action, cost-effective system improvements can be achieved while delivering more reliable electric service to customers.

Courtesy of National Grid.

National Grid
Waltham, Massachusetts, USA
By Jeff Pires

Contact
Jeff Pires
jeffrey.pires@us.ngrid.com

Software
ArcGIS Desktop 9.2, ArcGIS Spatial Analyst

Printer
HP Designjet 1050c Plus

Data Source
National Grid

Geo-CEMI: Spatial Analysis of Customer-Level Electricity Reliability

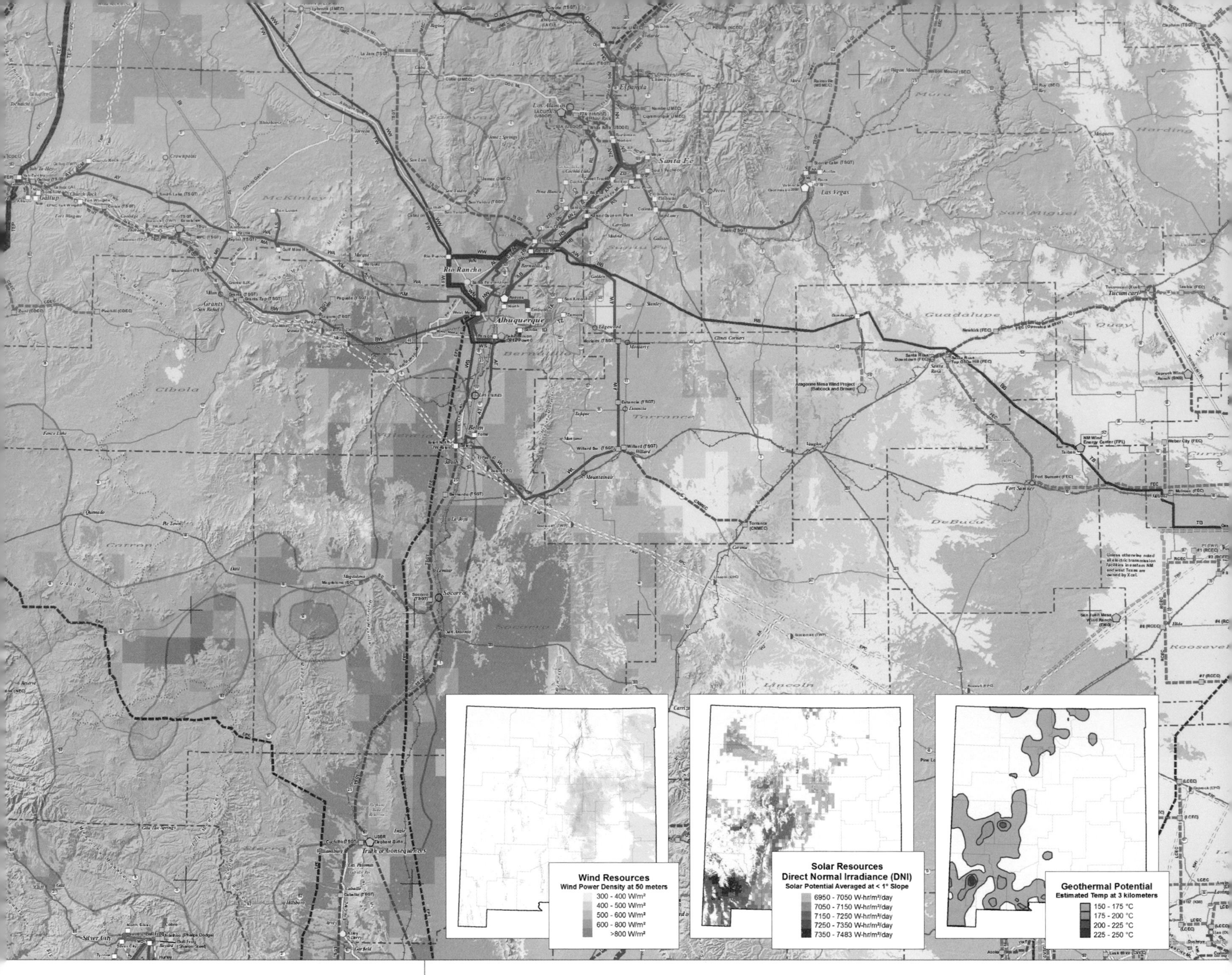

Public Service Company of New Mexico (PNM)
Albuquerque, New Mexico, USA
By Larry Rose, Douglas Campbell, and John Evaskovich

Contact
Douglas Campbell
zapper@swcp.com

Software
ArcGIS Desktop 9.2

Printer
HP Designjet 5000

Data Sources
True Wind Solutions Idaho National Lab, National Renewable Energy Lab, Southern Methodist University, basemap from PNM, U.S. Geological Survey, U.S. Census Bureau, Electric Cooperative System Maps

PNM is New Mexico's largest electricity and natural gas provider and is based in Albuquerque. PNM developed this map showing the relationship of wind, solar, and geothermal energy sources to existing electric transmission facilities for use in discussions with various state and federal agencies and the public.

The map illustrates both the limited overlap of the three types of renewable energy potential and the need for more electric transmission lines if renewable resources are to be developed further in New Mexico. Smaller maps show the geographic variations in renewable energy potential.

Courtesy of PNM.

New Mexico Renewable Energy Resource Potential with Existing Energy Transmission Lines

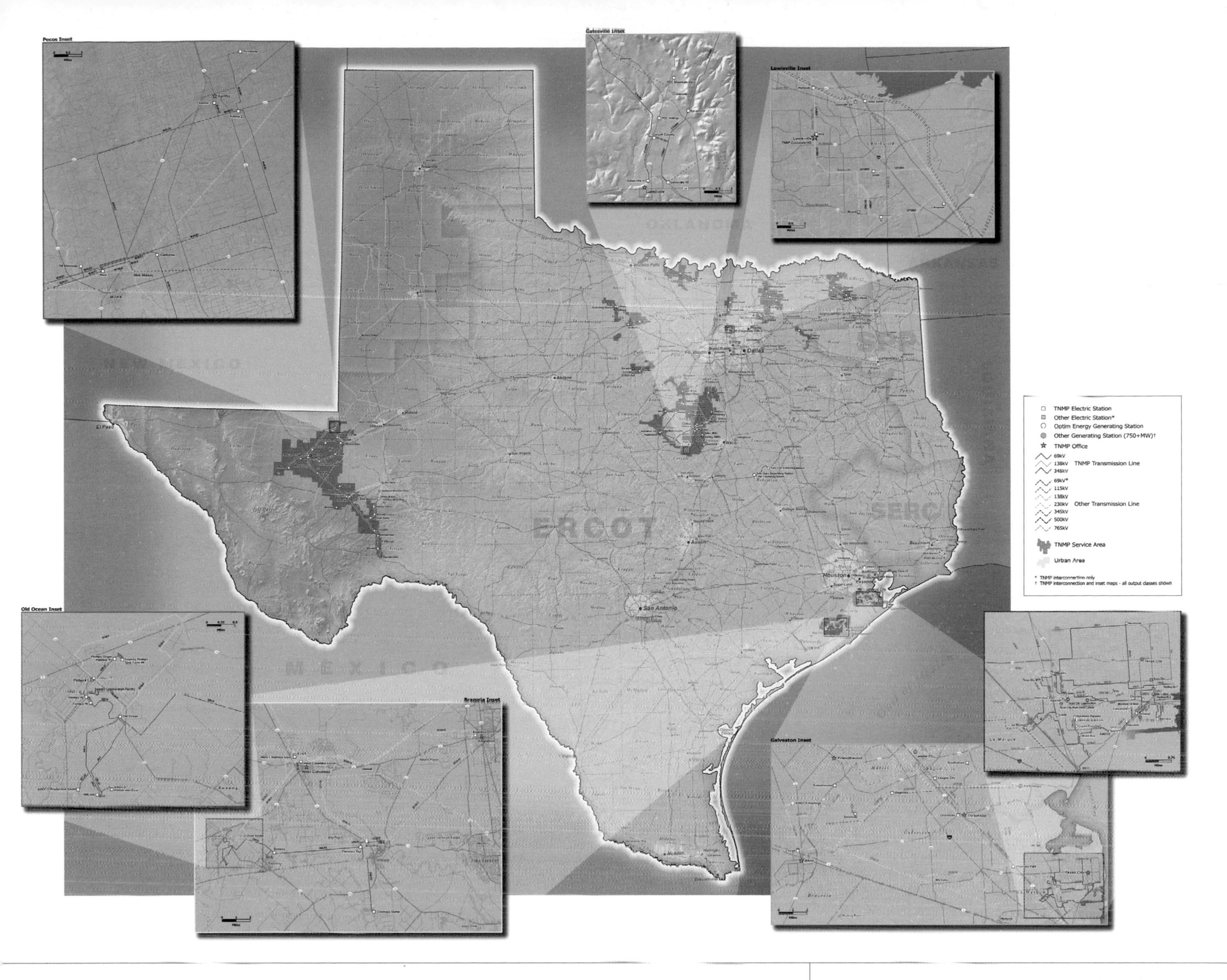

This map illustrates TNMP's transmission system and its relation to the electrical transmission grid in Texas. TNMP, formerly Texas-New Mexico Power Company, is a relatively small utility primarily focused on providing retail electric service. Its transmission lines support its distribution system (indicated by the service area polygons).

TNMP does not operate any large, high voltage intrastate transmission lines. In addition to TNMP's infrastructure being totally lost in the vast complex sea of the Texas grid, TNMP operations are dispersed throughout the state sometimes in areas of very small geographic extent and with very densely packed facilities. This cartographic challenge was met with an extensive use of inset maps and ArcGIS cartographic feature-class representations. This map is intended for a wide audience within TNMP and throughout the electric utility industry in Texas.

TNMP has provided community-based electric service since 1935 and currently supports seventy-six towns and more than 216,000 customers throughout Texas. The utility is owned by PNM Resources, an energy holding company based in Albuquerque, New Mexico.

Courtesy of PNM Resources.

Public Service Company of New Mexico (PNM)
Albuquerque, New Mexico USA
By John Evaskovich

Contact
John Evaskovich
John.Evaskovich@pnm.com

Software
ArcGIS Desktop, Adobe Photoshop

Printer
HP Designjet 5000ps

Data Sources
TNMP, Ventyx (formerly Global Energy), ESRI StreetMap, Electrical Reliability Council of Texas, North American Electric Reliability Corp, U.S .Geological Survey

TNMP Electric Transmission System

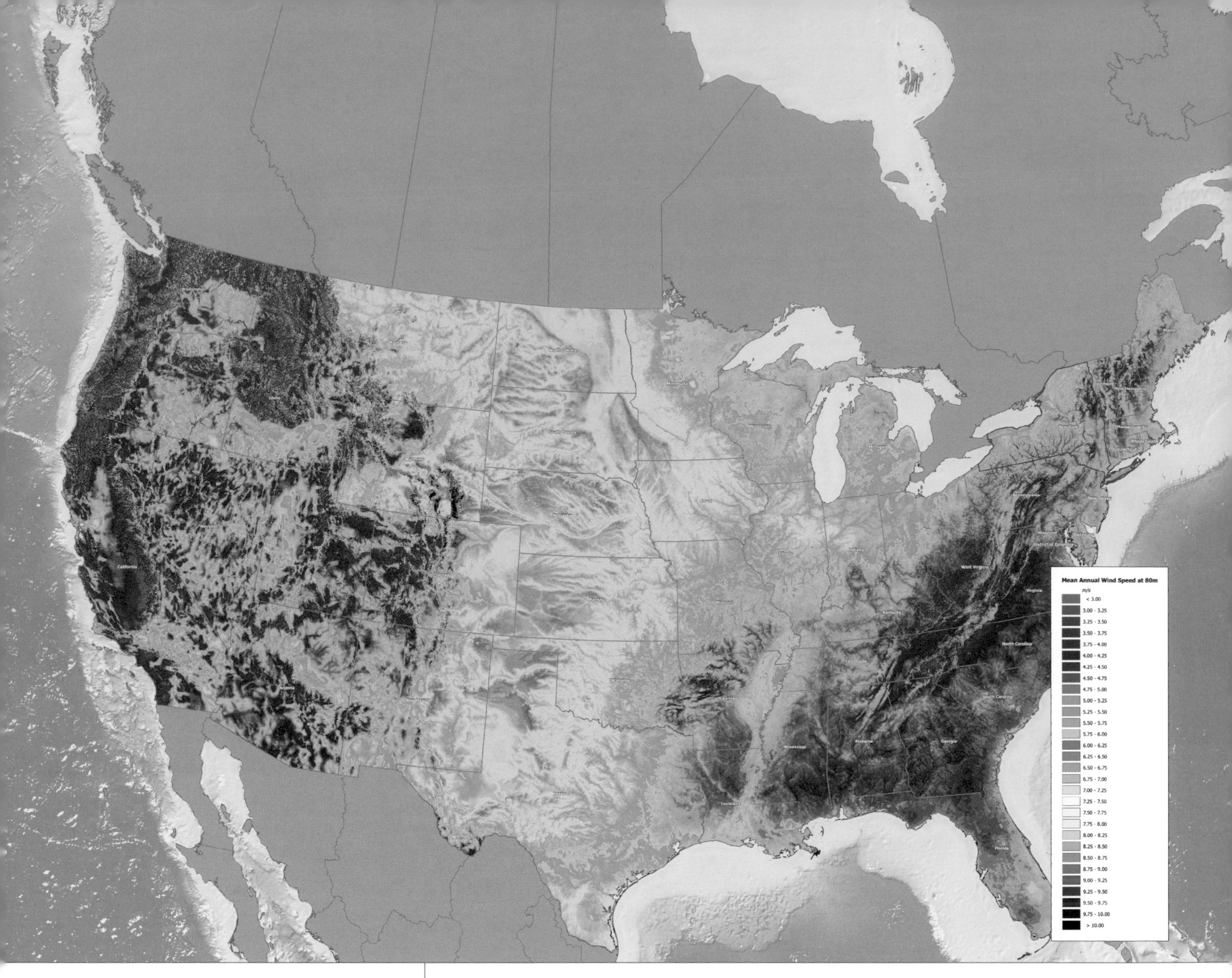

AWS Truewind, LLC
Albany, New York, USA
By Staci Clark and Erik Hale

Contact
Staci Clark
sclark@awstruewind.com

Software
ArcGIS Desktop, ArcGIS Spatial Analyst, ArcScene

Printer
HP Designjet 800ps 42

Data Sources
AWS 200 m wind map, Global Energy Decisions (Ventyx) transmission, NLCD 01

This image displays AWS Truewind's first national map for mean annual wind speed at 80 meters above ground. The wind resource data is the basis for identifying potential land area for wind development. Although AWS Truewind uses various methods for screening sites, the example depicted in this map for New York state features a simple approach using some basic GIS layers.

This study used a series of raster-based analyses at 200m resolution looking at land-cover type, topography, wind resource, transmission availability, and minimum land requirements. By aggregating small regions within a reasonable distance of one another, sizeable chunks of land were identified for large-scale wind development. The result was an estimate of total megawatt potential based on land area availability that will be used to make policy decisions and plan for future wind development in the state.

AWS Truewind, LLC, is a firm based in Albany, New York, that specializes in renewable energy consulting services to developers, investors, and governmental and institutional clients around the world.

Developing Wind Farms: Screening for Potential Sites

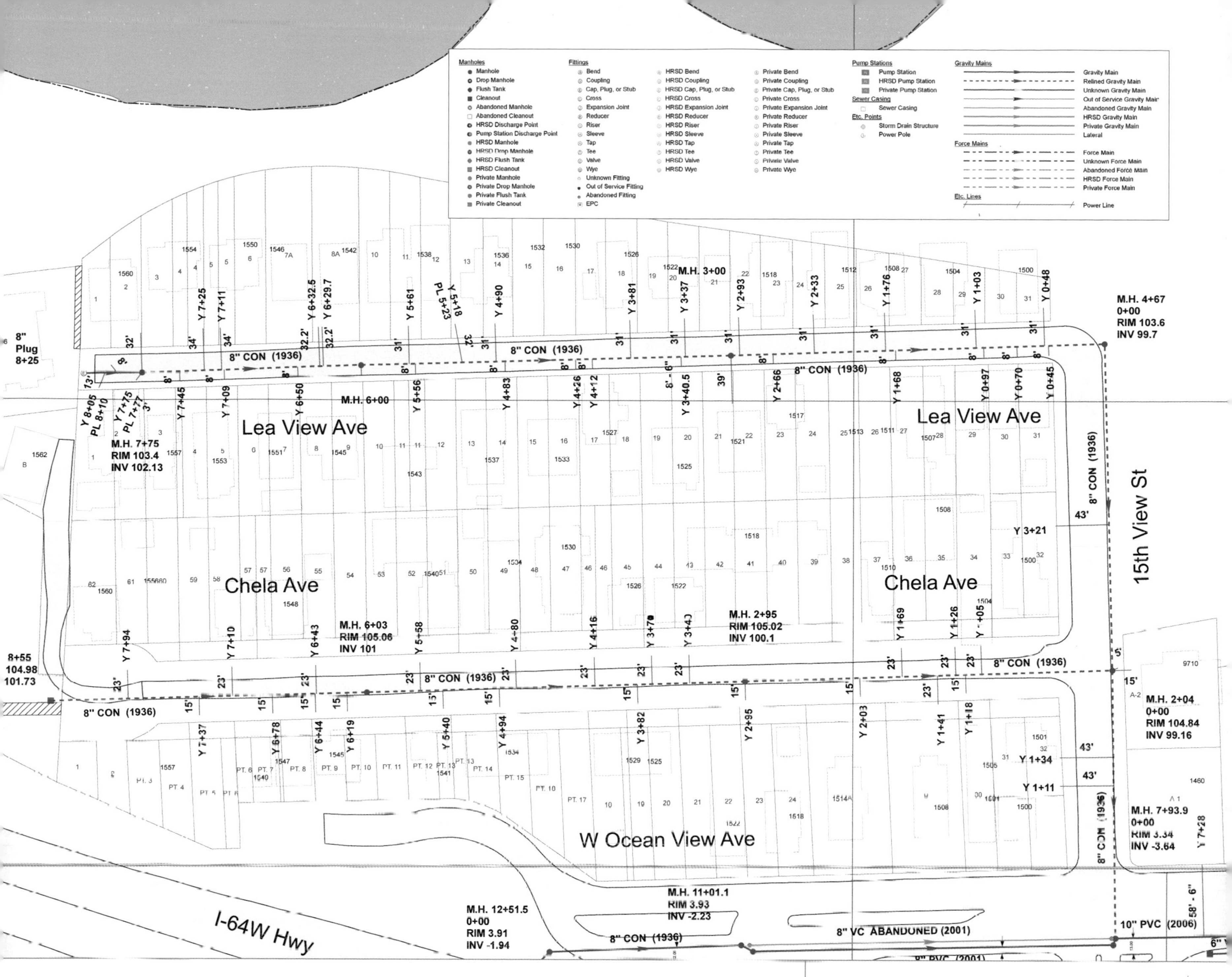

This is the design for the physical maps located in the City of Norfolk's Department of Utilities buildings. They are used for planning and emergency operations purposes. The map was created using ArcGIS Desktop 9.2 with the Map Book Developer Sample.

The Norfolk Department of Utilities was established on July 1, 1969, and was charged with the operation of the city's water and sewer systems. The department currently provides water for more than 800,000 people in the Hampton Roads area and maintains over 1,600 miles of water and sewer infrastructure. The current hand-drawn/computer-aided design record system is being replaced by GIS data. This process involves the GIS team researching the old record system and combining that information with new information that comes in daily from current infrastructure improvement and repair projects in the city.

Two types of maps were made: planimetrics (or large overview maps) and quads (used for more detailed views). The quad map is shown here. These maps are generally used for customer service purposes, such as locating sewer availability for residents, and also serve as emergency backups to computerized maps. The maps are designed for a 24-by-36-inch sheet, with a scale of 1:600 (1 inch equals 50 feet), and have a 1.5-inch overlap (eliminating those instances with a line appearing on four separate sheets). The maps keep the same detail and information as the old drawings; however, using GIS provides more up-to-date information, simplified updating, modeling capabilities, and the ability to sync maps with work order tracking systems.

Courtesy of City of Norfolk, Virginia, Department of Utilities.

City of Norfolk, Department of Utilities
Norfolk, Virginia, USA
By Nathaniel Davis, GISP and Alex English

Contact
Nathaniel Davis
ndavis@norfolk.gov

Software
ArcGIS Desktop

Printer
HP Designjet 4500mfp

Data Source
City of Norfolk

Sewer Quad D-01-E

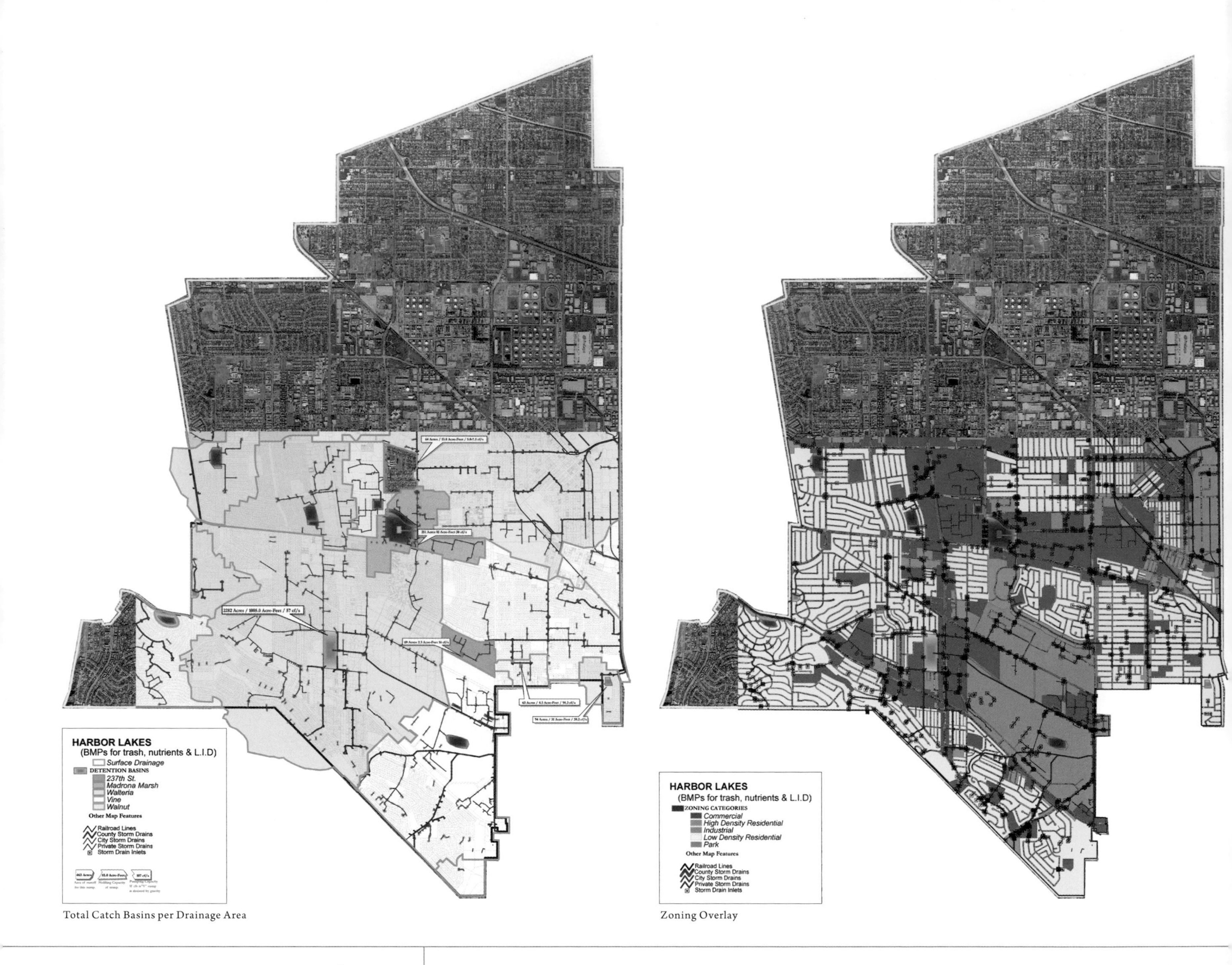

Total Catch Basins per Drainage Area

Zoning Overlay

City of Torrance
Torrance, California, USA
By Jennifer Gough, Sunny Lai, and John Dettle

Contact
Jennifer Gough
jgough@torrnet.com

Software
ArcGIS Desktop 9.2, Adobe Photoshop CS3

Printer
HP Designjet 5500ps

Data Sources
City of Torrance, Los Angeles Region Imagery Acquisition Consortium

These maps were created for Machado Lake's trash total maximum daily load monitoring and reporting plans, and form part of the City of Torrance best management practices for the National Pollutant Discharge Elimination System. One map scenario used detention basins for trash filtration and elimination from the system. The other map setting supported a proposal to install inserts in all catch basins, where trash quantities and filtration needs are related to broad land-use categories. The available GIS tools helped determine drainage acreages and number of catch basins in each drainage area for extraction into reports.

Torrance Storm Drain Mitigation Maps

In order to ensure the city's sanitary sewer system is well-maintained, Santa Monica city inspectors routinely inspect sewer mains using a truck-mounted camera. Incorporating Granite XP software into this process allows quick, user-friendly data entry to document pipe failures or confirm lateral locations and condition.

City of Santa Monica
Santa Monica, California, USA
By Wesley Cantu and Michael A. Carson

Contact
Michael A. Carson
michael.carson@smgov.net

Software
ArcGIS Desktop 9.2, Granite XP

Printer
HP Designjet 5500 ps

Data Sources
Santa Monica GIS, Santa Monica Water Resources

Index by Organization

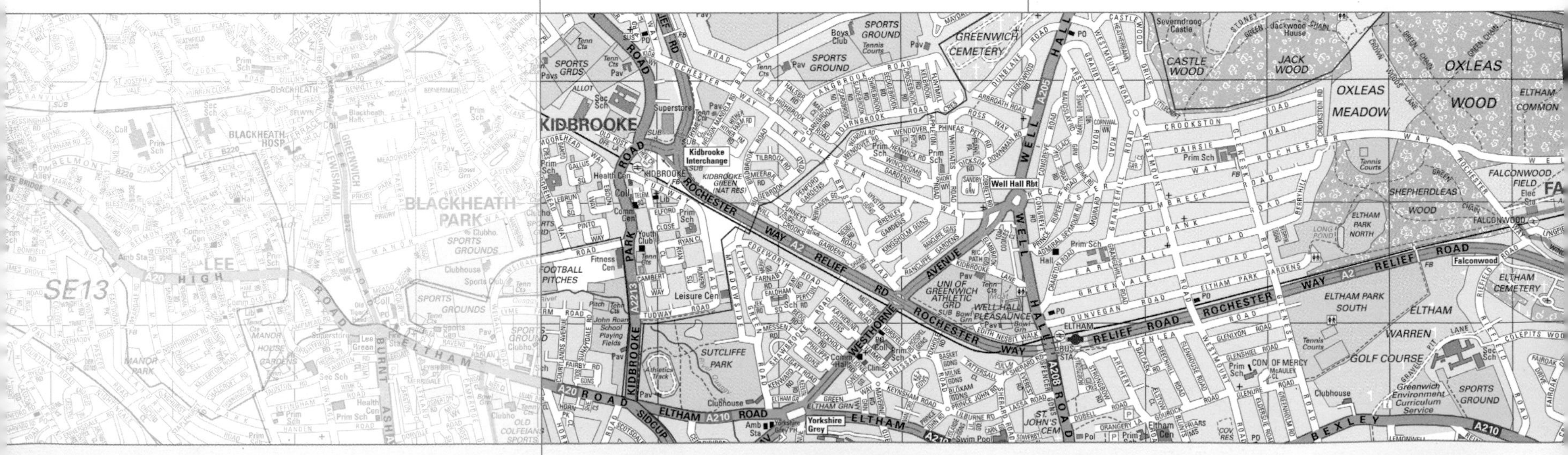